AF576037

Processing and Characterization of Magnesium-Based Materials

Processing and Characterization of Magnesium-Based Materials

Editor

Domonkos Tolnai

MDPI • Basel • Beijing • Wuhan • Barcelona • Belgrade • Manchester • Tokyo • Cluj • Tianjin

Editor
Domonkos Tolnai
MagIC - Magnesium Innovation Center
Germany

Editorial Office
MDPI
St. Alban-Anlage 66
4052 Basel, Switzerland

This is a reprint of articles from the Special Issue published online in the open access journal *Crystals* (ISSN 2073-4352) (available at: https://www.mdpi.com/journal/crystals/special_issues/magnesium-based_materials).

For citation purposes, cite each article independently as indicated on the article page online and as indicated below:

LastName, A.A.; LastName, B.B.; LastName, C.C. Article Title. *Journal Name* **Year**, *Volume Number*, Page Range.

ISBN 978-3-0365-1102-3 (Hbk)
ISBN 978-3-0365-1103-0 (PDF)

Contents

About the Editor vii

Preface to "Processing and Characterization of Magnesium-Based Materials" ix

Domonkos Tolnai
Processing and Characterization of Magnesium-Based Materials
Reprinted from: *Crystals* **2021**, *11*, 96, doi:10.3390/cryst11020096 1

Gerardo Garces, Rafael Barea, Andreas Stark and Norbert Schell
Anisotropic Plastic Behavior in an Extruded Long-Period Ordered Structure$Mg_{90}Y_{6.5}Ni_{3.5}$ (at.%) Alloy
Reprinted from: *Crystals* **2020**, *10*, 279, doi:10.3390/cryst10040279 3

Daria Drozdenko, Gergely Farkas, Pavol Šimko, Klaudia Fekete, Jan Čapek, Gerardo Garcés, Dong Ma, Ke An and Kristián Máthis
Influence of Volume Fraction of Long-Period Stacking Ordered Structure Phase on the Deformation Processes during Cyclic Deformation of Mg-Y-Zn Alloys
Reprinted from: *Crystals* **2020**, *11*, 11 , doi:10.3390/cryst11010011 17

Ricardo Henrique Buzolin, Leandro Henrique Moreno Guimaraes, Julián Arnaldo Ávila Díaz, Erenilton Pereira da Silva, Domonkos Tolnai, Chamini L. Mendis, Norbert Hort and Haroldo Cavalcanti Pinto
Restoration Mechanisms at Moderate Temperatures for As-Cast ZK40 Magnesium Alloys Modified with Individual Ca and Gd Additions
Reprinted from: *Crystals* **2020**, *10*, 1140, doi:10.3390/ccryst10121140 31

Thorsten Henseler, Shmuel Osovski, Madlen Ullmann, Rudolf Kawalla and Ulrich Prahl
GTN Model-Based Material Parameters of AZ31 Magnesium Sheet at Various Temperatures by Means of SEM In-Situ Testing
Reprinted from: *Crystals* **2020**, *10*, 856, doi:10.3390/cryst10100856 53

Wenyan Zhang, Hua Zhang, Lifei Wang, Jianfeng Fan, Xia Li, Lilong Zhu, Shuying Chen, Hans Jørgen Roven and Shangzhou Zhang
Microstructure Evolution and Mechanical Properties of AZ31 Magnesium Alloy Sheets Prepared by Low-Speed Extrusion with Different Temperature
Reprinted from: *Crystals* **2020**, *10*, 644, doi:10.3390/cryst10080644 71

Kamineni Pitcheswara Rao, Kalidass Suresh, Yellapregada Venkata Rama Krishna Prasad and Manoj Gupta
Thermomechanical Processing of AZ31-3Ca Alloy Prepared by Disintegrated Melt Deposition (DMD)
Reprinted from: *Crystals* **2020**, *10*, 647, doi:10.3390/cryst10080647 81

Hui Su, Zhibing Chu, Chun Xue, Yugui Li and Lifeng Ma
Relationship among Initial Texture, Deformation Mechanism, Mechanical Properties, and Texture Evolution during Uniaxial Compression of AZ31 Magnesium Alloy
Reprinted from: *Crystals* **2020**, *10*, 738, doi:10.3390/cryst10090738 95

About the Editor

Domonkos Tolnai is a research scientist in the Institute of Metallic Biomaterials at the Helmholtz-Zentrum Geesthacht (Geesthacht, Germany). He obtained his PhD degree in Materials Science at the Vienna University of Technology (Vienna, Austria) in 2011. From 2007 to 2011, he worked on the in situ synchrotron tomographic characterization of Al–Mg–Si alloys during solidification. Since 2011m his research has focused on the microstructure-property relationship of Mg-based metallic materials and their response to thermo–mechanical load during processing with in situ synchrotron–diffraction and –imaging techniques. He is the author and/or co-author of one book chapter, more than 50 peer-reviewed publications in international journals, and more than 40 talks on international scientific conferences.

Preface to "Processing and Characterization of Magnesium-Based Materials"

As one of the lightest structural materials, magnesium has by now substituted heavier counterparts in many applications. However, it possesses sound-specific mechanical properties, poor corrosion resistance, and moderate absolute properties, which impede its widespread application. Conscious alloying and surface treatment can improve the corrosion resistance, and when the degradation rate is controlled, certain Mg alloys can be used as bio-resorbable implant materials. Improving the mechanical properties of magnesium has been the focus of numerous research studies. The addition of alloying elements and refining the processing routes allows tailoring the microstructure and consequently the properties to fit the specific application. One example is the simultaneous addition of zinc and rare earth elements that leads to the formation of a long-period stacking-ordered (LPSO) structure under favorable processing conditions, resulting in a significant improvement in strength. These investigations and the resulting data advance the computational modeling approaches to construct a digital counterpart of the material, which allows a faster and more resource-efficient alloy development in the future.

Domonkos Tolnai

Editor

crystals

MDPI

Editorial

Processing and Characterization of Magnesium-Based Materials

Domonkos Tolnai

Institute of Metallic Biomaterials, Helmholtz-Zentrum Geesthacht, Max-Planck-Straße 1, D 21502 Geesthacht, Germany; Domonkos.Tolnai@hzg.de; Tel.: +49-415-287-1974

Mg-based materials have become increasingly attractive for industries, where weight saving is of importance (e.g., automotive, aerospace). Their low density and good specific properties qualify them to substitute heavier materials. Their macroscopic property profile, however, needs to be improved in order to meet all the requirements these applications demand.

One approach to improve the mechanical properties of Mg is to introduce a secondary phase with a reinforcement capacity by the addition of alloying elements. The combination of transition metals and rare-earth elements under favorable processing conditions leads to the formation of long-period stacking order (LPSO) structures, providing outstanding room- and elevated-temperature strength. Garcés et al. [1] investigated the deformation behavior of an extruded Mg90Y6.5Ni3.5 alloy with an almost fully LPSO structure at different temperatures by means of in situ synchrotron radiation diffraction during compression. The results show an anisotropy in strength between the extrusion and the transversal direction. This could be explained by the orientation dependence at the activation of basal slip. When basal slip is hindered, kinking becomes the main deformation mechanism. At elevated temperature, where more deformation mechanisms could be activated, this anisotropy decreases.

check for updates

Citation: Tolnai, D. Processing and Characterization of Magnesium-Based Materials. *Crystals* **2021**, *11* , 96 . https://doi.org/10.3390/cryst11020096

Received: 19 January 2021
Accepted: 20 January 2021
Published: 22 January 2021

Drozdenko et al. [2] also investigated the deformation behavior of Mg (WZ) alloys containing an LPSO structure with in situ neutron diffraction during cyclic loading. The study found that in the alloy with the lowest volume fraction of LPSO, the twinning–detwinning mechanism found in non-DRX α-Mg grains affects the overall deformation. On the other hand, the increase in the volume fraction of LPSO allows an increased load transfer into the second phase. Therefore, slip and kinking of LPSO dominate the plastic deformation and the twinning–detwinning mechanism only plays a minor role.

In situ diffraction is a unique experimental technique to follow the changes in the different phases in the microstructure during deformation and to determine the main operating deformation mechanism under load. This technique was utilized by Buzolin et al. [3] when investigating the deformation in ZK40 alloys with the addition of Ca and Gd during elevated-temperature compression. The results show that besides dislocation slip, twinning, dynamic recrystallization and dynamic recovery take place, and the latter is the predominant mechanism. The addition of Ca and Gd impedes dynamic recrystallization compared to the ZK40 without alloying additions.

Besides alloying, the microstructure design and consequently the design of the property profile can be conducted by optimizing the processing of the material as well. Henseler et al. [4] investigated the deformation properties of AZ31 sheets at different temperatures produced via different processing routes by in situ SEM investigations. The authors obtained experimental values of parameters used in the Gurson–Tvergaard–Needleman (GTN) model to be further used in finite element simulations.

Similar to the latter study, Zhang et al. [5] investigated the microstructure evolution and the mechanical properties of AZ31 sheets prepared by low-speed extrusion between 350 and 450 °C. The results show that the resulting grain size of the extruded material decreases, while the maximal basal texture intensity decreases with the increase in temperature. Extrusion at 350 °C yielded the best overall mechanical properties.

AZ31 modified with up to 3 wt.% Ca was prepared by disintegrated melt deposition and investigated by Rao et al. [6] to obtain processing maps. Three domains could be identified on the processing map for AZ31-3Ca, where in Domain I, basal and prismatic slip associated with dynamic recovery; in Domain II, grain boundary sliding promoted intercrystalline cracking; and in Domain III, dynamic recrystallization by non-basal slip and recovery by climb occurred.

The effect of the initial texture on the deformation behavior of AZ31 was investigated by Su et al. [7] with EBSD and visco-plastic self-consistent (VPSC) modeling. The study shows how the initial texture plays a major role in the texture evolution during mechanical loading, leading to mechanical anisotropy.

This Special Issue provides insight into the forefront of magnesium research where the improvement in the mechanical property profile along with the corrosion resistance is of great importance to promote the widespread application of Mg-based materials.

Conflicts of Interest: The author declare no conflict of interest.

References

1. Garcés, G.; Barea, R.; Stark, A.; Schell, N. Anisotropic Plastic Behavior in an Extruded Long-Period Ordered Structure Mg90Y6.5Ni3.5 (at.%) Alloy. *Crystals* **2020**, *10*, 279. [CrossRef]
2. Drozdenko, D.; Farkas, G.; Šimko, P.; Fekete, K.; Čapek, J.; Garcés, G.; Ma, D.; An, K.; Máthis, K. Influence of Volume Fraction of Long-Period Stacking Ordered Structure Phase on the Deformation Processes during Cyclic Deformation of Mg-Y-Zn Alloys. *Crystals* **2021**, *11*, 11. [CrossRef]
3. Buzolin, R.H.; Guimaraes, L.H.M.; Díaz, J.A.Á.; da Silva, E.P.; Tolnai, D.; Mendis, C.L.; Hort, N.; Pinto, H.C. Restoration Mechanisms at Moderate Temperatures for As-Cast ZK40 Magnesium Alloys Modified with Individual Ca and Gd Additions. *Crystals* **2020**, *10*, 1140. [CrossRef]
4. Henseler, T.; Osovski, S.; Ullmann, M.; Kawalla, R.; Prahl, U. GTN Model-Based Material Parameters of AZ31 Magnesium Sheet at Various Temperatures by Means of SEM In-Situ Testing. *Crystals* **2020**, *10*, 856. [CrossRef]
5. Zhang, W.; Zhang, H.; Wang, L.; Fan, J.; Li, X.; Zhu, L.; Chen, S.; Roven, H.J.; Zhang, S.-Z. Microstructure Evolution and Mechanical Properties of AZ31 Magnesium Alloy Sheets Prepared by Low-Speed Extrusion with Different Temperature. *Crystals* **2020**, *10*, 644. [CrossRef]
6. Rao, K.; Suresh, K.; Prasad, Y.V.R.K.; Gupta, M. Thermomechanical Processing of AZ31-3Ca Alloy Prepared by Disintegrated Melt Deposition (DMD). *Crystals* **2020**, *10*, 647. [CrossRef]
7. Su, H.; Chu, Z.; Xue, C.; Li, Y.; Ma, L. Relationship among Initial Texture, Deformation Mechanism, Mechanical Properties, and Texture Evolution during Uniaxial Compression of AZ31 Magnesium Alloy. *Crystals* **2020**, *10*, 738. [CrossRef]

Article

Anisotropic Plastic Behavior in an Extruded Long-Period Ordered Structure $Mg_{90}Y_{6.5}Ni_{3.5}$ (at.%) Alloy

Gerardo Garces [1,*], Rafael Barea [2], Andreas Stark [3] and Norbert Schell [4]

1. Department of Physical Metallurgy, CENIM-CSIC, Avenida Gregorio del Amo 8, 28040 Madrid, Spain
2. Departamento de Ingeniería Industrial, Universidad Nebrija, Campus Dehesa de la Villa, C. Pirineos 55, 28040 Madrid, Spain; rbarea@nebrija.es
3. Institute of Materials Research, Helmholtz-Zentrum Geesthacht, Max-Planck-Str. 1, 21502 Geesthacht, Germany; Andreas.Stark@hzg.de
4. Structural Research on New Materials, Helmholtz- Zentrum Geesthacht Outstation at DESY, Notkestraße 85, 22607 Hamburg, Germany; norbert.schell@hzg.de

* Correspondence: ggarces@cenim.csic.es; Tel.: +34-91-553-8900; Fax: +34-91-534-7425

Received: 12 March 2020; Accepted: 2 April 2020; Published: 7 April 2020

Abstract: The $Mg_{90}Y_{6.5}Ni_{3.5}$ alloy composed almost completely of the Long-Period-Stacking-Ordered (LPSO) phase has been prepared by casting and extrusion at high temperature. An elongated microstructure is obtained where the LPSO phase with 18R crystal structure is oriented with its basal plane parallel to the extrusion direction. Islands of α-magnesium are located between the LPSO grains. The mechanical properties of the alloy are highly anisotropic and depend on the stress sign as well as the relative orientation between the stress and the extrusion axes. The alloy is stronger when it is compressed along the extrusion direction. Under this configuration, the slip of <a> dislocations in the basal plane is highly limited. However, the activation of kinking induces an increase in the plastic deformation. In the transversal extrusion direction, some grains deform by the activation of basal slip. The difference in the yield stress between the different stress configurations decreases with the increase in the test temperature. The evolution of internal strains obtained during in-situ compressive experiments reveals that tensile twinning is not activated in the LPSO phase.

Keywords: magnesium alloys; long period stacking ordered structures (LPSO); synchrotron radiation diffraction

1. Introduction

Mg-Transition metal (TM: Zn, Ni, Cu, Co and Al)-Rare Earth (RE: Y, Gd, Dy, Ho, Er, etc.) alloys with a Long-Period Stacking Ordered (LPSO) crystal structure have attracted great attention due to their high mechanical strength (around 600 MPa), which is maintained up to 300 °C. The LPSO phases are typically Mg-TM-RE compounds whose structure show a long range stacking of basal, hexagonal planes with periodic enrichment of TM and RE atoms in basal planes [1–14]. LPSO crystal structures can be represented by $L1_2$ type ordered TM_6RE_8 clusters embedded in local fcc stacking layers [15–18].

The mechanical properties of the LPSO phase have been widely studied in the Mg-Zn-Y system. The LPSO phase with 18R or 14H structure shows a strong elastic and plastic anisotropy. In the elastic regime, the LPSO phase exhibits a higher Young's modulus and shear modulus compared to Mg [19–22] and depending on the LPSO crystal orientation. On the other hand, its plastic behavior is controlled by the activation of the basal slip system with a CRSS of around 10 MPa [23,24]. During extrusion, the LPSO phase is oriented with their basal plane parallel to the extrusion direction. The activation of the basal slip system is forbidden when the material is tested along the extrusion direction, and the LPSO phase exhibits differences in the mechanical behavior depending on the stress sign and the orientation of the stress [25–27]. Under compression along the extrusion direction, the material exhibits the highest yield stress (500–550 MPa). Moreover, the activation of kinking [23–27] and non-basal slip systems [28] induces a high plastic deformation capacity. Under tension along the extrusion direction, the LPSO phase is extremely brittle.

Itoi et al. [29] have reported that rolled-Mg-Y-Ni alloy with almost fully LPSO phase exhibited a high tensile mechanical strength (460 MPa) at room temperature with a ductility of 8%. This alloy with these high mechanical strength values can substitute aluminum or steels for some structural applications. However, the mechanical behavior under different stress configuration has not been explored to analyze their plastic anisotropy. The present study examines the orientation dependence of strength of an extruded $Mg_{90}Y_{6.5}Ni_{3.5}$ alloy with an almost fully LPSO phase and its temperature dependences. The deformation mechanisms occurring during plastic deformation have been analyzed using in-situ synchrotron radiation diffraction experiments at room and high temperatures during compression tests in a direction parallel and perpendicular to the extrusion axis.

2. Materials and Methods

The alloy with a nominal composition of 90%Mg-6.5%Y-3.5%Ni (atomic percent) was prepared by melting in an electric resistance furnace using high purity Mg and Ni elements and a Mg-22%Y master alloy. Ingots were cast by pouring the liquid metal into a cylindrical steel mold of diameter 42 mm. The alloy was homogenized at 350 °C for 24 h and then extruded at 450 °C using an extrusion ratio of 18:1 (a rectangular die of section 7 mm by 12 mm) and an extrusion rate of 0.5 mm/s.

Microstructure was examined by scanning (SEM) (JEOL JSM 6500F, JEOL, Tokyo, Japan) and transmission (TEM) (JEOL JEM 2010, JEOL, Tokyo, Japan) electron microscopy. Samples for SEM were prepared by mechanical polishing and finishing with an etching solution of 5 mL acetic acid, 20 mL water and 25 mL picric acid in methanol. Specimens for TEM were prepared by ion milling at liquid nitrogen temperature.

For mechanical testing, samples with different shape and size were prepared from the extruded bar and machined for loading along the extrusion direction (ED) and the longest transversal direction (TD). For compression tests at different temperatures, prismatic samples of $5 \times 5 \times 10$ mm^3 were prepared. For tensile tests along ED direction, cylindrical samples with a diameter of 3 mm and length 10 mm were used. Tensile and compressive tests were carried out in air at temperatures between room temperature and 400 °C, using a strain rate of 4×10^{-4} s^{-1}. In compression, tests were stopped at about 10% of strain if the sample was not broken before.

Synchrotron radiation diffraction (SRD) was carried out at the P07–HEMS beamline of PETRA III, at the Deutsches Elektronen-Synchrotron (DESY, Hamburg, Germany) during in-situ compression tests to identify the deformation mechanisms of the $Mg_{90}Y_{6.5}Ni_{3.5}$ alloy. The diffraction patterns were recorded in a fast mode using an exposure time of around 0.5 s by a Perkin-Elmer XRD 1621 flat-panel detector with an array of 2048×2048 pixels^2, with an effective pixel size of 200×200 µm^2. The beam energy and, therefore, the wavelength were 100 keV and 0.0124 nm, respectively. As a reference, LaB_6 was used to calibrate the system. The detector-to-sample distance was 1789 mm. The in-situ compressive test at a strain rate of 10^{-3} s^{-1} where carried out in a DIL 805A/D (TA Instruments, New Castle, DE, USA) dilatometer at room temperature and 300 °C. Cylindrical samples of 5 mm of diameter and 10 mm of length where machined from the extruded bar along the extrusion and the longest transversal direction. The thermocouple was welded at the surface of the compressive sample to control the test temperature. The beam was positioned at the centre of the sample with the gauge volume defined approximately by the primary slits (1×1 mm^2) and the cylinder diameter, which allows a good grain statistic.

2θ diffraction profiles were calculated by azimuthal integration of a 10° section (±5°) of the Debye-Scherrer rings in the axial and radial direction, i.e., parallel and perpendicular to the compression direction. The fitting of each individual diffraction peak was carried out using the FIT2D software (European Synchrotron Radiation Facility (ESRF), Grenoble, France) [30] and a Gaussian function. The elastic strain for each diffraction peak is calculated by the relative shift in the position of the diffraction peak:

$$\varepsilon_{hkl} = \frac{d_{hkl} - d_{0,hkl}}{d_{0,hkl}} \tag{1}$$

where d_{hkl} and $d_{0,hkl}$ are the interplanar distance of the hkl plane in the stressed and stress-free crystal. $d_{0,hkl}$ is selected as the interplanar distance before the compression test. The diffraction angle θ and the lattice planar spacing are linked through the Bragg's law.

$$d_{hkl} = \frac{\lambda}{2\sin\theta_{hkl}} \tag{2}$$

where λ is the wavelength of the radiation.

3. Results

The 3D microstructure of the extruded material, shown in Figure 1a, was characterized by the presence of two phases, which are elongated along the extrusion direction. The preponderance phase (light gray phase) corresponds to the LPSO while the dark phase corresponds to alfa magnesium with some yttrium and nickel in solid solution. The volume fraction of the darker Mg phase, calculated using image analysis in the as-cast condition, covered about 13.4% ± 0.8% and was located at the grain boundaries of LPSO grains. TEM and SRD studies were carried out to determine the crystallographic structure of the LPSO phase and its crystallographic texture. Figure 1b,c shows the bright-field image for the LPSO phase using g = 0002 diffraction vector as well as the selected area electron diffraction (SAED) pattern in the $[11\bar{2}0]$ zone axis. Fringes, perpendicular to the [0002] direction, were observed. The fringe spacing was 1.6 nm. The SAED pattern of the LPSO in the $[11\bar{2}0]$ zone axis presents six diffracted spots along the [0002] directions between the transmitted spot and the (0002) diffracted spot of the magnesium structure. Both observations confirmed the 18R (rhombohedral) crystal structure of the LPSO phase after extrusion.

Figure 2a shows the Debye–Scherrer rings obtained in the extruded $Mg_{90}Y_{6.5}Ni_{3.5}$ alloy extruded at 450 °C. After integration along the extrusion and transversal directions, diffraction patterns as a function of 2θ are obtained (Figure 2b). The crystal structure of the LPSO phase corresponds to the 18R structure ($P3_212$ space group) with parameters a = 1.11 nm and c = 4.69 nm. The diffraction peaks with highest intensities in the transversal direction correspond to $(0003)_{18R}$ and $(00018)_{18R}$, which are not observed in the diffraction pattern obtained along the extrusion direction. During the extrusion process, the LPSO phase reorients its basal plane parallel to the extrusion direction. Therefore, the alloy exhibited a strong fiber texture with the basal plane parallel to the extrusion direction and it is expected that mechanical properties of the extruded $Mg_{90}Y_{6.5}Ni_{3.5}$ alloy show differences when tested along the extrusion or transversal directions.

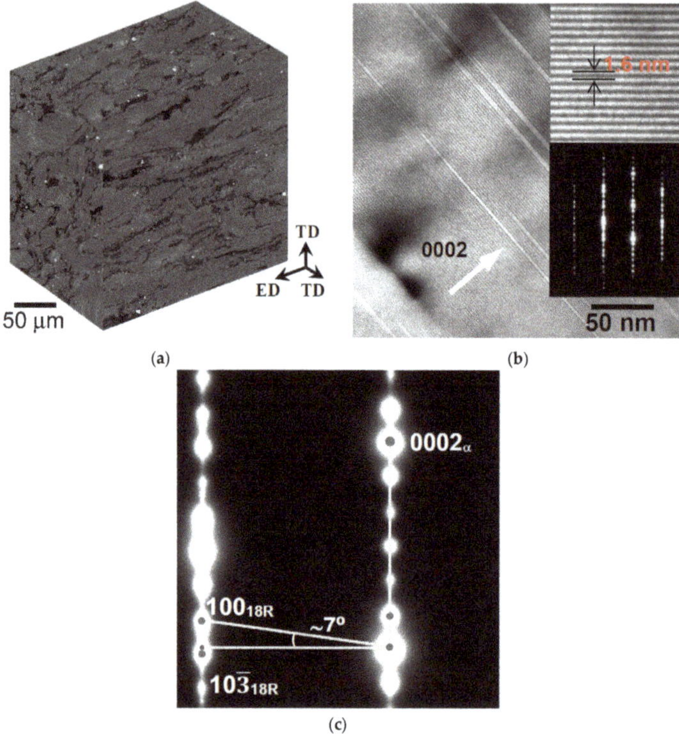

Figure 1. (**a**) 3D-SEM micrographs showing microstructure of extruded $Mg_{90}Y_{6.5}Ni_{3.5}$ alloy. (**b**) TEM micrograph showing the Long-Period Stacking Ordered (LPSO) grains (B = [$11\bar{2}0$], g = (0002)) and its corresponding SAED. A detail of the fringes is also shown. The diffraction pattern in (**c**) is an enlarged part of the SAED in Figure 1b for clarity.

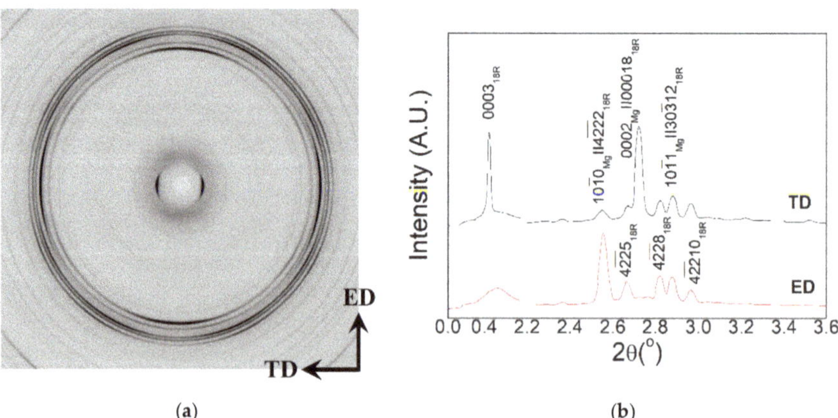

Figure 2. (**a**) Synchrotron diffraction pattern recorded on the 2D flat-panel detector after diffraction by the extruded $Mg_{90}Y_{6.5}Ni_{3.5}$ alloy. (**b**) Diffraction pattern in the extrusion direction (ED) and longest transversal direction (TD) directions as a function of 2θ obtained by integration of the data in (a).

True stress–true strain curves of the extruded $Mg_{90}Y_{6.5}Ni_{3.5}$ alloy tested in tension and in compression from room temperature (RT) to 400 °C are shown in Figure 3a–c. The shape of the curves was essentially the same at a given temperature for each sample orientation independently of tension and compression. The work hardening was high in the beginning of the tensile or compressive curve (about 1–2% strain), but then decreased to low values. Failure in tension occurred at low strain at low temperature (1% at room temperature). Tensile ductility improved considerably above 100 °C. Figure 3d shows yield stresses (flow stress at 0.2% plastic strain) determined from the stress–strain curves of Figure 3a–c. In general terms, all materials show high strength at low temperatures and fall to very low stresses above 200 °C.

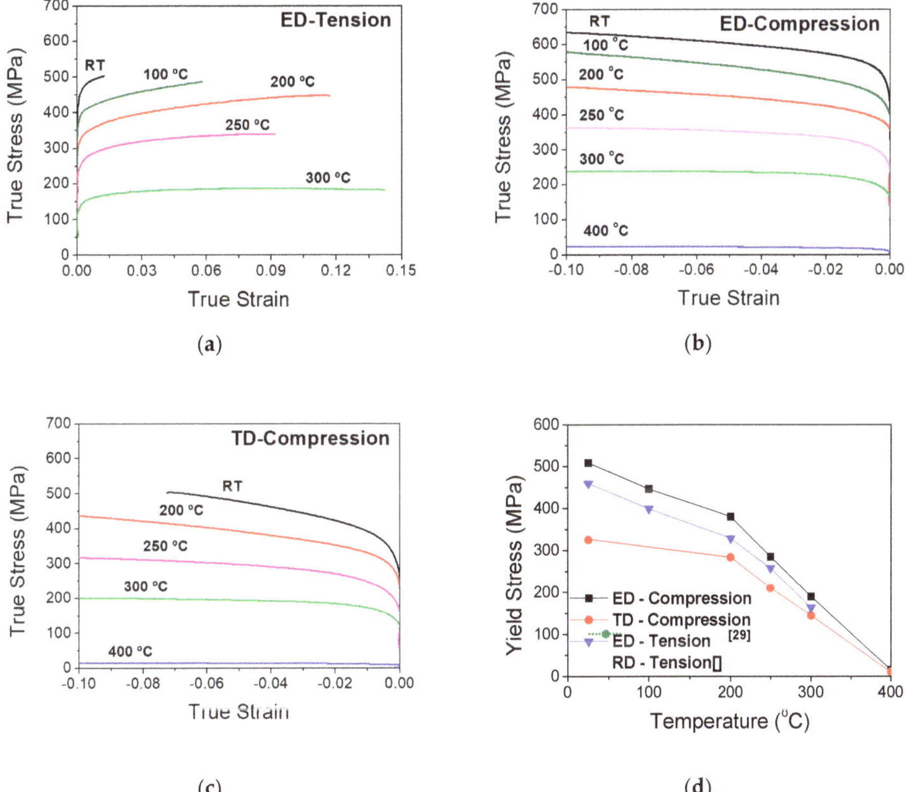

Figure 3. True stress–strain curves obtained from room temperature to 400 °C (**a**) in tension along the extrusion direction, (**b**) compression along the extrusion direction and (**c**) compression along the longest transversal direction. (**d**) Yield stress variation with test temperature for materials tested in tension and in compression for samples strained along extrusion and transverse directions. RD is referred to the rolling direction from yield stress values obtained from reference [29].

Yield stresses of samples tested in the extrusion direction were the highest, being some 180 MPa and 130 MPa (compression and tension tests, respectively) higher than samples compressed in the transversal direction at room temperature. This behavior was maintained up to 400 °C. However, the stress difference between different deformation modes decreased with the increase in the test temperature. For the orientation ED tested in tension, the strength was 50 MPa lower than in compression at room temperature. The stress difference between tensile and compressive test along the extrusion direction also decreased with the increase in the test temperature.

The tension–compression asymmetry of yield stress in the extruded $Mg_{90}Y_{6.5}Ni_{3.5}$, with strength in compression shown to be significantly greater than strength in tension, has been previously reported in the Mg-Y-Zn alloys [26] and it is opposite to that observed in extruded magnesium alloys. SRD experiments during compressive tests at room temperature and 300 °C have been carried out to study in detail the influence of the test temperature on the mechanical anisotropic behavior. Figure 4 shows the integrated synchrotron diffraction patterns in axial direction, (obtained from images similar to Figure 2a), before (σ = 0 MPa) and after the compression tests (ε around 8–10%) for both loading orientations (ED and TD) at RT and 300 °C, respectively. In ED, the diffraction peak with the highest intensity, due to the strong fiber texture described in the above sections, corresponds to $\{4\bar{2}\bar{2}2\}_{18R}$ planes ($\{10\bar{1}0\}_{Mg}$ planes in the magnesium hexagonal crystal structure) perpendicular to the axial direction. After compression (red line), diffraction peaks were shifted to higher values of 2θ (lower d values). Moreover, the intensity of diffraction peaks slightly decreased and the peak width highly increased. On the other hand, in TD, the diffraction peak with the highest intensity in the axial direction corresponded to $\{00018\}_{18R}$ planes perpendicular to the axial direction. Similar to ED, diffraction peaks were shifted to higher 2θ values and they were broader compared to initial diffraction peaks before the compression test. The intensity of diffraction peaks did not change significantly. Diffraction patterns obtained at 300 °C follow the same behavior as room temperature. However, the peak shift and peak width for a similar plastic strain were lower than at room temperature.

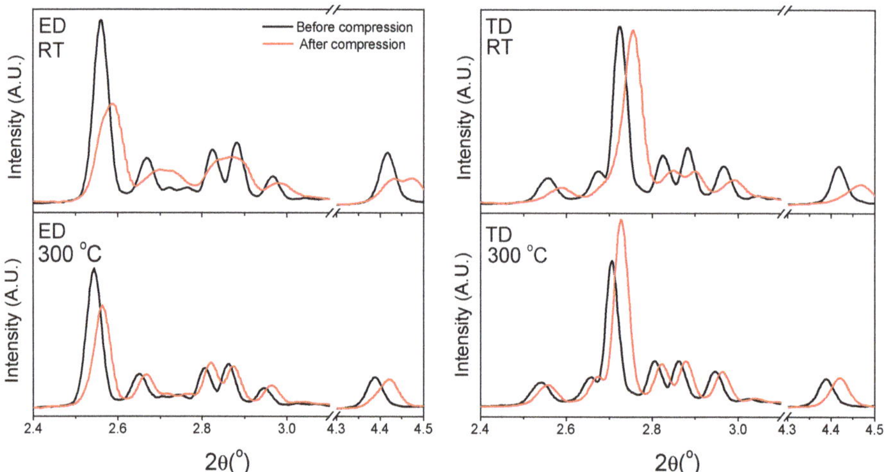

Figure 4. Synchrotron axial diffraction patterns as a function of 2θ before and after compression test at room temperature and 300 °C in the ED and TD directions. Each graph defines the condition: ED-RT (Extrusion direction, room temperature), TD-RT (Transverse direction, room temperature), ED-300 °C (Extrusion direction, 300 °C) and TD-300 °C (Transverse direction, 300 °C)

Figure 5b,c,f,g shows the evolution of the elastic strains in the axial and radial direction, obtained from Equation (1), as a function of the applied stress for the $\{4\bar{2}\bar{2}2\}_{18R}(\{10\bar{1}0\}_{Mg})$, $\{00018\}_{18R}$ ((0002_{Mg}), $\{4\bar{2}\bar{2}8\}_{18R}$, $\{30\bar{3}12\}_{18R}$ ($\{10\bar{1}1\}_{Mg}$) and $\{60\bar{6}0\}_{18R}$ ($\{11\bar{2}0\}_{Mg}$) diffraction peaks of the 18R structure at room temperature and 300 °C. The compressive curves for both cases were also plotted for comparison purposes (Figure 5a,e). At room temperature, the macroscopic yield stress was 510 and 320 MPa in ED and TD, respectively. The stress difference between both directions was in agreement with compressive yield stress data shown in Figure 3. In ED at room temperature, diffraction peaks show a linear elastic behavior in the elastic regime except for the $\{30\bar{3}12\}_{18R}$ diffraction peak that lost its linearity at 350 MPa. This diffraction peak in the rhombohedral 18R crystal structure was equivalent to the $\{10\bar{1}1\}_{Mg}$ diffraction peak in the hexagonal structure of the magnesium phase. Although, the extruded alloy exhibited a strong fiber texture with the basal plane parallel to the extrusion direction, there were also grains with other orientations, which were favorable oriented to the activate the basal slip system. However, since their volume fraction is small, they had no influence in the macroscopic mechanical behavior. At yield stress, the elastic strain corresponding to the $\{4\bar{2}\bar{2}2\}_{18R}$ diffraction peak in the axial direction begin to lose its linearity (marked with and arrow in Figure 5b). This effect was accompanied with the rapid increase in the elastic strain, higher than in the elastic slope, of the $\{00018\}_{18R}$ diffraction peak in the radial direction (around 2000 μstrains) while the internal strain of all other diffraction peaks remained constant.

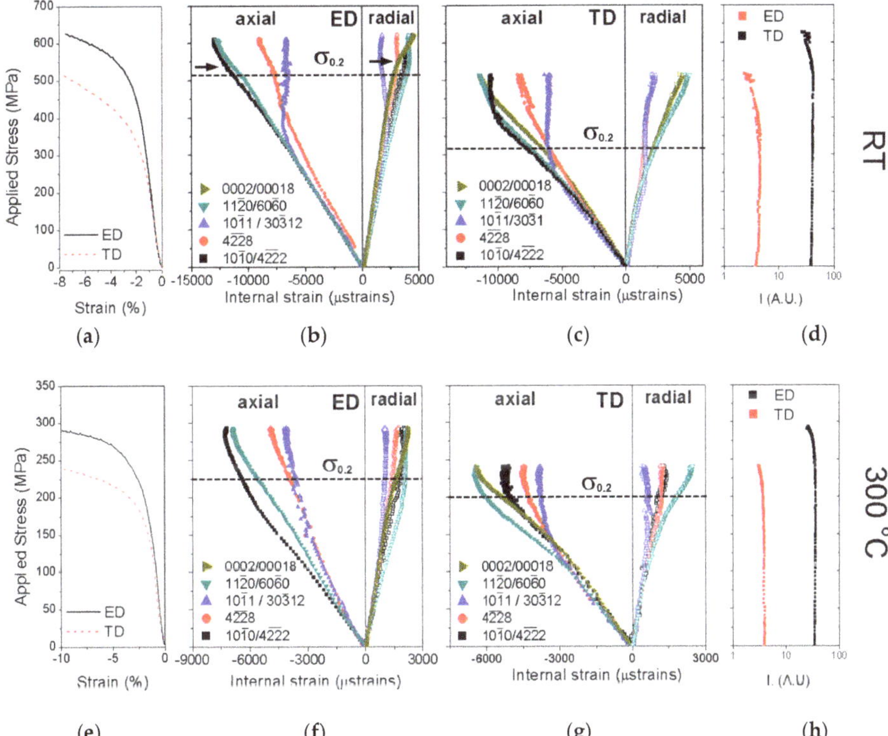

Figure 5. Compressive macroscopic stress–strain curves of the extruded $Mg_{90}Y_{6.5}Ni_{3.5}$ alloy obtained during the in-situ experiment for the sample deformed along extrusion and transversal directions for (**a**) room temperature and (**e**) 300 °C. Axial and radial internal strains for the $\{4\bar{2}\bar{2}2\}_{18R}(\{10\bar{1}0\}_{Mg})$, $\{00018\}_{18R}((0002)_{Mg})$, $\{4\bar{2}\bar{2}8\}_{18R}$, $\{30\bar{3}12\}_{18R}(\{10\bar{1}1\}_{Mg})$ and $\{60\bar{6}0\}_{18R}(\{11\bar{2}0\}_{Mg})$ diffraction peaks for the 18R structure as a function of the applied stress during an in-situ compression test along (**b**,**f**) extrusion and (**c**,**g**) transversal directions at (**b**,**c**) room temperature and (**f**,**g**) 300 °C. The dash line represents the yield stress in the in-situ compressive curve. Evolution of the integrated intensity of $\{4\bar{2}\bar{2}2\}_{18R}(\{10\bar{1}0\}_{Mg})$ diffraction peaks in the axial direction for the extruded alloy during compression along extrusion and transversal direction at (**d**) room temperature and (**h**) 300 °C.

At higher temperature, Figure 5f, the analysis was more complex since the elastic strains of almost all diffraction peaks lost their linearity at different applied stresses due to the activation of different deformation systems. In ED at 300 °C, the yield stress was around 225 MPa, which was connected with the progressive loss of linearity of the $\{4\overline{2}\overline{2}2\}_{18R}$ diffraction peak, similar to the compression test at room temperature.

The evolution of the elastic strains as a function of the applied stress in the sample tested along TD at room temperature was similar to the ED condition, but changes in the internal strain of each family of grains took place at lower applied stress values. The elastic strains of all diffraction peaks (axial and radial directions) show a linear elastic behavior below the macroscopic yield stress except for the $\{30\overline{3}12\}_{18R}$ that lost its linearity at 250 MPa. The elastic strain corresponding to the $\{4\overline{2}\overline{2}2\}_{18R}$ diffraction peak in the axial direction lost its linearity at higher stress than the macroscopic yield stress. Therefore, the macroscopic yield stress was nearest to the loss of linearity of $\{30\overline{3}12\}_{18R}$ that corresponds to the slip of <a> dislocations in the basal plane. It is important to point out that in TD, the $\{00018\}_{18R}$ diffraction peak exhibited the highest integrated intensity in the axial direction and, therefore, the evolution of the elastic strain could be calculated. The elastic strain remained elastic up to 450 MPa. At 300 °C, the behavior was again similar to room temperature (Figure 5g) and only internal strain changes occurred at lower applied stress values.

Magnesium alloys developed a strong fiber texture during the extrusion process with the basal plane parallel to the extrusion direction, similar to that observed in the $Mg_{90}Y_{6.5}Ni_{3.5}$ alloy. Under compression along the extrusion direction, extension twinning should be activated. Extension twinning induced a rotation of the basal plane, which was oriented almost perpendicular to the compression axis after plastic deformation. This crystal rotation results in a strong decrease in the intensity of the $\{10\overline{1}0\}_{Mg}$ diffraction peak. Figure 5d and h show the evolution of the integrated intensity in logarithmic scale of the $\{4\overline{2}\overline{2}2\}_{18R}(\{10\overline{1}0\}_{Mg})$ as a function of the applied stress in the axial direction. The intensity was almost constant or slightly decreased when the internal strain of the $\{4\overline{2}\overline{2}2\}_{18R}$ diffraction peak lost its linearity. Therefore, extension twinning seemed not to occur during compression along the extrusion and transversal directions.

4. Discussion

The microstructure of the $Mg_{90}Y_{6.5}Ni_{3.5}$ was characterized by 18R-LPSO coarse grains elongated along the extrusion direction and oriented with their basal plane parallel to the extrusion direction. This microstructure was in agreement with previous studies in a similar alloy after hot rolling [29]. In both cases, islands of magnesium elongated along the extrusion direction are observed at LPSO boundaries. Magnesium islands are solidified previous to the solidification of the LPSO phase in the as-cast condition due to their higher melting point compared to the LPSO phase [31].

Several crystal structures have been reported for the LPSO phases in the Mg-Ni-Y system. Depending on the Ni and Y content and thermal treatment of the alloy 10H, 12R, 14H, 18R and 24R were reported [29,32–38]. Wang et al. [34] studied the Mg-Ni-Y equilibrium diagram using a thermodynamic calculation. They proposed that the 18R structure was stable during the cast and transforms to 14H at around 535 °C. Moreover, the 18R structure could also transform to 10H trough a solid-state reaction at around 450 °C. On the other hand, Jiang et al. [35] have reported that the 14H structure is thermodynamically stable in the Mg-Ni-Y system. In the MgY_2Zn_1 alloy, the 18R crystal structure transformed to 14H (hexagonal) after a thermal treatment at temperatures higher than 300 °C. The synchrotron diffraction pattern allowed distinguishing univocally between the different crystal structures at low diffraction angles. The (0003) diffracted peak of the 18R structure ($P3_212$, a = 1.11 nm and 4.69 nm [15]) appeared at 0.45°, the (0002) diffracted peak of the 14H structure (P63/mcm, a = 1.11 nm and c = 3.62 nm [15]) appeared at 0.39°and the (0002) diffracted peak of the 10H structure (P63/mcm, a = 1.11 nm and c = 2.60 nm [39]) at 0.55°. Figure 2b clearly showed that only 18R structure was present after the extrusion at 450 °C. Therefore, the extrusion process was not sufficient to transform the 18R structure. The 18R lattice parameter values correspond very closely to those obtained for the 18R structure of the cast Mg-Y-Zn [31].

The alloy composition and the microstructure of the extruded bar examined here was essentially similar to that reported earlier by Itoi et al. [29] on hot rolled alloy. Even, the mechanical yield stress in the rolling direction was similar to the tensile behavior in ED (Figure 3d). Although, the hot rolled alloy exhibited higher elongation than the extruded alloy at room temperature.

The main deformation mechanism reported in the 18R structure in the Mg-Y-Zn system was the slip of <a> dislocations in the basal plane. The critical resolved shear stress is around 10 MPa [23]. Figure 6 shows a scheme of the microstructure of the extruded bar and the texture of the elongated LPSO grains with the basal plane parallel to the extrusion direction. In the $Mg_{90}Y_{6.5}Ni_{3.5}$ alloy, elongated grains were oriented with their basal planes parallel to the extrusion direction. Therefore, the slip of <a> dislocations in the basal plane was inhibited under the load along the extrusion direction. This fact would result in a loss of ductility of the extruded $Mg_{90}Y_{6.5}Ni_{3.5}$ alloy not only in tension but also in compression. However, the compression test along the extrusion direction was stopped at a plastic strain of 0.1 without sample failure. The microstructural characterization of the polished surface of the compressive sample after deformation exhibited a high volume fraction of kinks (Figure 7). The formation of kink bands is connected with the presence of basal dislocations [23]. Mayama et al. [40] has demonstrated using crystal plasticity simulation that small fluctuation (around 3°) from the perfect orientation between the [0002] and compression axis results in a strong decrease in the onset of plasticity due to the activation of kinking. The loss of linearity of the $\{4\overline{2}\overline{2}2\}_{18R}$ diffraction peak was, therefore, related with the activation of kinking. The formation of a kink band induced an elastic strain along the radial direction in the basal plane of the LPSO phase. At 300 °C, the plasticity was still controlled by the activation of kinking. However, the stress decreased almost by 275 MPa.

On the other hand, in the TD condition, there was a volume fraction of grains that were well-oriented for the activation of the basal slip and, therefore, yield stress highly decreased. However, there were still grains oriented with the basal plane parallel or perpendicular to the compression axis where the slip of <a> dislocations in the basal plane was forbidden and, therefore, kinking was activated. At 300 °C, the applied stress for the activation of kinking was the same as in the ED mode. Therefore, the decrease of the yield stress in TD was caused by the higher decrease in the CRSS of the basal slip system at 300 °C.

The evolution of internal strains and intensities of diffraction peaks shown in Figure 5 seemed to reveal that tensile twinning did not occur during deformation of the extruded $Mg_{90}Y_{6.5}Ni_{3.5}$ alloy. The activation of the tensile twinning in magnesium alloys presented two main characteristics. On one hand, tensile twinning induced a rotation of 86° of the basal plane within the twin area. Therefore, the intensity of the $(0002)_{Mg}$ diffraction peak increased during the compression test at the expense of the $\{10\overline{1}0\}_{Mg}$ diffraction peak due to the formation of twins. On the other hand, the internal strain as a function of the applied stress of twins oriented with the $(0002)_{Mg}$ planes perpendicular to the compression axis highly increased faster than in the elastic regime. The appearance of the diffraction peak corresponding to the basal plane, {00018} in the 18R crystal structure, was not observed in the axial direction during the complete compression test, as it was observed in Figure 5. In contract, the diffraction peaks were broader compared to the initial width due to kinking. Kinking induced the local rotation (±10°) of the basal planes between kink bands and initial grains.

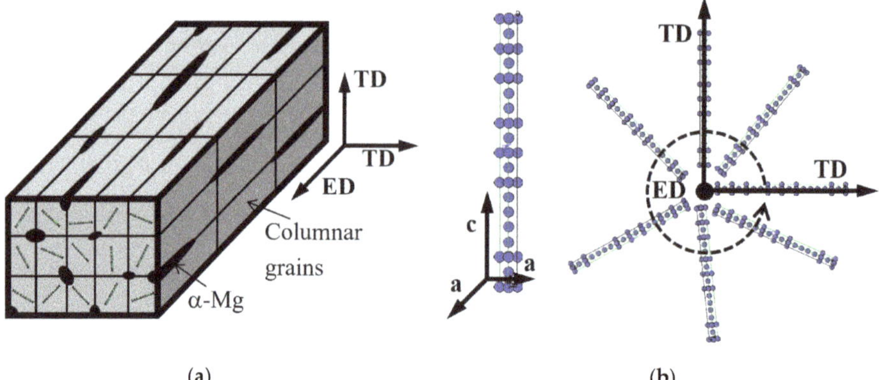

(a) (b)

Figure 6. (**a**) Scheme of the microstructure of the alloy with columnar LPSO grains and elongated magnesium islands along the extrusion direction. (**b**) LPSO lattice and texture of the crystal lattice in the columnar grains.

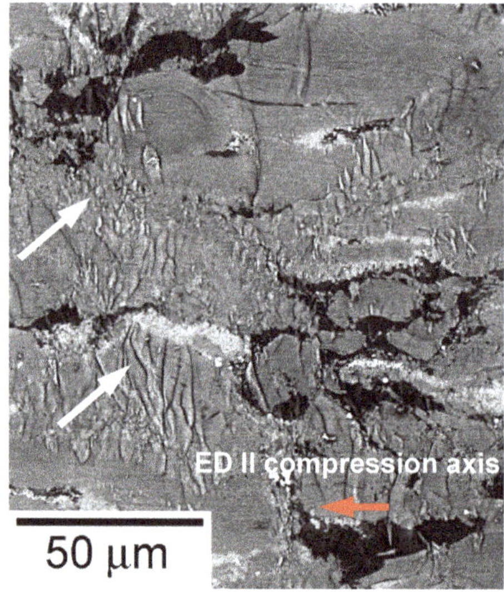

Figure 7. SEM micrographs showing surface features on polished samples strained about 10% in compression at room temperature along the extrusion direction. Kinks are marked in the images with white arrows.

5. Conclusions

The plastic anisotropy in an $Mg_{90}Y_{6.5}Ni_{3.5}$ alloy extruded at high temperature was studied in a wide temperature range. The microstructure was composed of elongated LPSO grains elongated along the extrusion direction with small amounts of Mg phase between the LPSO grains. A strong texture was built up during extrusion, whereby basal planes were arranged strictly parallel to the extrusion axis. The following conclusions can be drawn:

1. The material was much stronger when tested in the extrusion direction, compared to material tested in the transversal direction. This orientation dependence might be explained by the ease of activation of the basal slip system according to the orientation of the basal planes in the samples.

2. When the orientation prevented the deformation in the basal plane, the deformation mechanism that was activated was the kinking. The stress for the activation of kinking decreased with the increase in the test temperature.

3. This yield stress differences between the loading directions decreased with the increase of the test temperature due to the decreases in the CRSS of other deformation systems, which promoted their activation.

Author Contributions: Conceptualization, G.G.; methodology, G.G. and R.B; validation, G.G., R.B., A.S. and N.S.; investigation, G.G. and R.B; writing—original draft preparation, G.G. and R.B.; writing—review and editing, G.G., R.B., A.S. and N.S. All authors have read and agreed to the published version of the manuscript.

Funding: This research was funded by SPANISH MINISTRY OF ECONOMY AND COMPETITIVENESS, grant number MAT2016-78850-R.

Acknowledgments: The Deutsches-Elektronen-Synchrotron DESY is acknowledged for the provision of beam time at the p07 beamline of the Petra III synchrotron radiation facility under the project I-20170054EC.

Conflicts of Interest: The authors declare no conflict of interest.

References

1. Kawamura, Y.; Yamasaki, M. Formation and mechanical properties of $Mg_{97}Zn_1RE_2$ alloys with long-period stacking ordered structure. *Mater. Trans.* **2007**, *48*, 2986–2992. [CrossRef]
2. Hagihara, K.; Kinoshita, A.; Sugino, Y.; Yamasaki, M.; Kawamura, Y.; Yasuda, H.Y. Effect of long-period stacking ordered phase on mechanical properties of $Mg_{97}Zn_1Y_2$ extruded alloy. *Acta Mater.* **2010**, *58*, 6282–6292. [CrossRef]
3. Yamasaki, M.; Sasaki, M.; Nishijima, M.; Hiraga, K.; Kawamura, Y. Formation of 14H long period stacking ordered structure and profuse stacking faults in Mg–Zn–Gd alloys during isothermal aging at high temperature. *Acta Mater.* **2007**, *55*, 6798–6805. [CrossRef]
4. Garcés, G.; Pérez, P.; González, S.; Adeva, P. Development of long-period ordered structures during crystallisation of amorphous $Mg_{80}Cu_{10}Y_{10}$ and $Mg_{83}Ni_9Y_8$. *Int. J. Mater. Res.* **2006**, *97*, 404–408.
5. Kishida, K.; Yokobayashi, H.; Inui, H.; Yamasaki, M.; Kawamura, Y. The crystal structure of the LPSO phase of the 14H-type in the Mg–Al–Gd alloy system. *Intermetallics* **2012**, *31*, 55–64. [CrossRef]
6. Yuan, L.T.; Bi, G.L.; Li, Y.D.; Jiang, J.; Han, Y.X.; Fang, D.Q.; Ma, Y. Effects of solid solution treatment and cooling on morphology of LPSO phase and precipitation hardening behaviour of Mg–Dy–Ni alloy. *Trans. Nonferrous Met. Soc. China* **2017**, *27*, 2381–2389. [CrossRef]
7. Xua, D.; Hana, E.H.; Xua, Y. Effect of long-period stacking ordered phase on microstructure, mechanical property and corrosion resistance of Mg alloys: A review. *Progress Nat. Sci. Mater. Inter.* **2016**, *26*, 117–128. [CrossRef]
8. Liu, J.; Yang, M.; Zhang, X.; Fan, D.; Che, C.; Zou, A. Effects of Ho content on microstructures and mechanical properties of Mg-Ho-Zn alloys. *Mater. Charact.* **2019**, *149*, 198–205. [CrossRef]
9. Zou, G.; Cai, X.; Fang, D.; Wang, Z.; Zhao, T.; Peng, Q. Age strengthening behavior and mechanical properties of Mg–Dy based alloys containing LPSO phases. *Mater. Sci. Eng. A* **2015**, *620*, 10–15. [CrossRef]
10. Bi, G.; Han, Y.; Jiang, J.; Li, Y.; Zhang, D.; Qiu, D.; Easton, M. Microstructure and mechanical properties of an extruded Mg-Dy-Ni alloy. *Mater. Sci. Eng. A* **2019**, *760*, 246–257. [CrossRef]
11. Leng, Z.; Zhang, J.; Yin, T.; Zhang, L.; Guo, X.; Peng, Q.; Zhang, M.; Wu, R. Influence of biocorrosion on microstructure and mechanical properties of deformed Mg–Y–Er–Zn biomaterial containing 18R-LPSO phase. *J. Mech. Behav. Biomed. Mater.* **2013**, *28*, 332–339. [CrossRef] [PubMed]
12. Peng, Z.Z.; Jin, Q.Q.; Shao, X.H.; Zhou, Y.T.; Zheng, S.J.; Zhang, B.; Ma, X.L. Effect of temperature on deformation mechanisms of the $Mg_{88}Co_5Y_7$ alloy during hot compression. *Mater. Charact.* **2019**, *151*, 553–562. [CrossRef]
13. Jin, Q.Q.; Shao, X.H.; Peng, Z.Z.; You, J.H.; Qiu, K.Q.; Ma, X.L. Crystallographic account of an ultra-long period stacking ordered phase in an $Mg_{88}Co_5Y_7$ alloy. *J. Alloys Compd.* **2017**, *693*, 1035–1038. [CrossRef]
14. Xiulan Su, M.J.; Li, H.; Ren, Y.; Qin, G. The phase equilibria and thermal stability of the long-period stacking ordered phase in the Mg–Cu–Y system. *J. Alloys Compd.* **2014**, *593*, 141–147.

15. Egusa, D.; Abe, E. The structure of long period stacking/order Mg–Zn–RE phases with extended non-stoichiometry ranges. *Acta Mater.* **2012**, *60*, 166–178. [CrossRef]
16. Nishioka, T.; Yamamoto, Y.; Kimura, K.; Hagihara, K.; Izuno, H.; Happo, N.; Hosokawa, S.; Abe, E.; Suzuki, M.; Matsushita, T.; et al. In-plane positional correlations among dopants in 10H type long period stacking ordered $Mg_{75}Zn_{10}Y_{15}$ alloy studied by X-ray fluorescence holography. *Materialia* **2018**, *3*, 256–259. [CrossRef]
17. Yamashita, K.; Itoi, T.; Yamasaki, M.; Kawamura, Y.; Abe, E. A novel long-period stacking/order structure in Mg-Ni-Y alloys. *J. Alloys Compd.* **2019**, *788*, 277–282. [CrossRef]
18. Egami, M.; Abe, E. Structure of a novel Mg-rich complex compound in Mg–Co–Y ternary alloys. *Scr. Mater.* **2015**, *98*, 64–67. [CrossRef]
19. Tane, M.; Kimizuka, H.; Hagihara, K.; Suzuki, S.; Nagai, Y. Effects of stacking sequence and short-range ordering of solute atoms on elastic properties of Mg–Zn–Y alloys with long-period stacking ordered structures. *Acta Mater.* **2015**, *96*, 170–188. [CrossRef]
20. Tane, M.; Nagai, Y.; Kimizuka, H.; Hagihara, K.; Kawamura, Y. Elastic properties of an Mg–Zn–Y alloy single crystal with a long-period stacking-ordered structure. *Acta Mater.* **2013**, *61*, 6338–6351. [CrossRef]
21. Garcés, G.; Máthis, K.; Medina, J.; Horváth, K.; Drozdenko, D.; Oñorbe, E.; Dobroň, P.; Pérez, P.; Klaus, M.; Adeva, P. Combination of in-situ diffraction experiments and acoustic emission testing to understand the compression behavior of Mg-Y-Zn alloys containing LPSO phase under different loading conditions. *Int. J. Plast.* **2018**, *106*, 107–128. [CrossRef]
22. Garces, G.; Pérez, P.; Cabeza, S.; Kabra, S.; Gan, W.; Adeva, P. Effect of Extrusion Temperature on the Plastic Deformation of an Mg-Y-Zn Alloy Containing LPSO Phase Using in-Situ Neutron Diffraction. *Metal. Mater. Trans. A* **2017**, *48*, 5332–5343. [CrossRef]
23. Hagihara, K.; Yokotani, N.; Umakoshi, Y. Plastic deformation behavior of $Mg_{12}YZn$ with 18R long-period stacking ordered structure. *Intermetallics* **2010**, *18*, 267–276. [CrossRef]
24. Hagihara, K.; Sugino, Y.; Fukusumi, Y.; Umakoshi, Y.; Nakamo, T. Plastic Deformation Behavior of $Mg_{12}ZnY$ LPSO-Phase with 14H-Typed Structure. *Mater. Trans.* **2011**, *52*, 1096–1103. [CrossRef]
25. Hagihara, K.; Yamasaki, M.; Kawamura, Y.; Nakano, T. Strengthening of Mg-based long-period stacking ordered (LPSO) phase with deformation kink bands. *Mater. Sci. Eng. A* **2019**, *763*, 138163. [CrossRef]
26. Garces, G.; Muñoz-Morris, M.A.; Morris, D.G.; Jimenez, J.A.; Perez, P.; Adeva, P. The role of extrusion texture on strength and its anisotropy in a Mg-base alloy composed of the Long-Period-Structural-Order phase. *Intermetallics* **2014**, *55*, 167–176. [CrossRef]
27. Hagihara, K.; Li, Z.; Yamasaki, M.; Kawamura, Y.; Nakano, T. Strengthening mechanisms acting in extruded Mg-based long-period stacking ordered (LPSO)-phase alloys. *Acta Mater.* **2019**, *163*, 226–239. [CrossRef]
28. Hagihara, K.; Fukusumi, Y.; Yamasaki, M.; Nakano, T.; Kawamura, Y. Non-Basal Slip Systems Operative in $Mg_{12}ZnY$ Long-Period Stacking Ordered (LPSO) Phase with 18R and 14H Structures. *Mater. Trans.* **2011**, *52*, 1096–1103. [CrossRef]
29. Itoi, T.; Takahashi, K.; Moriyama, H.; Hirohashi, M. A high-strength Mg–Ni–Y alloy sheet with a long-period ordered phase prepared by hot-rolling. *Scr. Mater.* **2008**, *59*, 1155–1158. [CrossRef]
30. Hammersley, A.P. *FIT2D: An Introduction and Overview*; ESRF Internal Report ESRF97HA02T; European Synchrotron Radiation Source: Grenoble, France, 1997.
31. Garcés, G.; Requena, G.; Tolnai, D.; Pérez, P.; Adeva, P.; Stark, A.; Schell, N. Influence of rare-earth addition on the long-period stacking ordered phase in cast Mg-Y-Zn alloys. *J. Mater. Sci.* **2014**, *49*, 2714–2722. [CrossRef]
32. Jin, Q.Q.; Fang, C.F.; Mi, S.B. Formation of long-period stacking ordered structures in $Mg_{88}M_5Y_7$ (M=Ti, Ni and Pb) casting alloys. *J. Alloys Compd.* **2013**, *568*, 21–25. [CrossRef]
33. Liu, C.; Zhu, Y.; Luo, Q.; Liu, B.; Gu, Q.; Li, Q. 12R long-period stacking odered structure in a Mg-Ni-Y alloy. *J. Mater. Sci. Tech.* **2018**, *34*, 2235–2239. [CrossRef]
34. Wang, Z.; Luo, Q.; Chen, S.; Chou, K.; Li, Q. Experimental investigation and thermodynamic calculation of the Mg–Ni–Y system (Y < 50 at.%) at 400 and 500 °C. *J. Alloys Compd.* **2015**, *649*, 1306–1314.
35. Jiang, M.; Zhang, S.; Bi, Y.; Li, H.; Ren, Y.; Qin, G. Phase equilibria of the long-period stacking ordered phase in the Mg-Ni-Y system. *Intermetallics* **2015**, *57*, 127–132. [CrossRef]
36. Zhang, Q.A.; Liu, D.D.; Wang, Q.Q.; Fang, F.; Sun, D.L.; Ouyang, L.Z.; Zhu, M. Superior hydrogen storage kinetics of $Mg_{12}YNi$ alloy with a long-period stacking ordered phase. *Scr. Mater.* **2011**, *65*, 233–236. [CrossRef]

37. Li, Y.; Gu, Q.; Lia, Q.; Zhang, T. In-situ synchrotron X-ray diffraction investigation on hydrogen-induced decomposition of long period stacking ordered structure in Mg–Ni–Y system. *Scr. Mater.* **2017**, *127*, 102–107. [CrossRef]
38. Zhu, S.M.; Lapovok, R.; Nie, J.F.; Estrin, Y.; Mathaudhu, S.N. Microstructure and mechanical properties of LPSO phase dominant $Mg_{85.8}Y_{7.1}Zn_{7.1}$ and $Mg_{85.8}Y_{7.1}Ni_{7.1}$ alloys. *Mater. Sci. Eng. A* **2017**, *692*, 35–42. [CrossRef]
39. Yamasaki, M.; Matsushita, M.; Hajihara, K.; Izuno, H.; Abe, E.; Kawamura, Y. Highly ordered 10H-type long-period stacking order phase in a Mg–Zn–Y ternary alloy. *Scr. Mater.* **2014**, *78–79*, 13–16. [CrossRef]
40. Mayama, T.; Ohashi, T.; Tadano, Y.; Hagihara, K. Crystal Plasticity Analysis of Development of Intragranular Misorientations due to Kinking in HCP Single Crystals Subjected to Uniaxial Compressive Loading. *Mater. Trans.* **2015**, *56*, 963–972. [CrossRef]

Article

Influence of Volume Fraction of Long-Period Stacking Ordered Structure Phase on the Deformation Processes during Cyclic Deformation of Mg-Y-Zn Alloys

Daria Drozdenko [1,2,*], Gergely Farkas [2], Pavol Šimko [1], Klaudia Fekete [1], Jan Čapek [3], Gerardo Garcés [4], Dong Ma [5], Ke An [6] and Kristián Máthis [1]

1 Department of Physics of Materials, Faculty of Mathematics and Physics, Charles University, Ke Karlovu 5, 121 16 Prague 2, Czech Republic; palko.simko@centrum.sk (P.Š.); fekete@karlov.mff.cuni.cz (K.F.); mathis@met.mff.cuni.cz (K.M.)

2 The Czech Academy of Sciences, Nuclear Physics Institute, Hlavní 130, 250 68 Řež, Czech Republic; farkasgr@gmail.com

3 Laboratory for Neutron Scattering and Imaging, Paul Scherrer Institute, 5232 Villigen PSI, Switzerland; jan.capek89@gmail.com

4 National Center for Metallurgical Research (CENIM-CSIC), Department of Physical Metallurgy, Avenida Gregorio del Amo 8, E-28040 Madrid, Spain; ggarces@cenim.csic.es

5 Neutron Science Platform, Songshan Lake Materials Laboratory, Dongguan 523808, China; dongma@sslab.org.cn

6 Oak Ridge National Laboratory (ORNL), 1 Bethel Valley Rd, Oak Ridge, TN 37830, USA; kean@ornl.gov

* Correspondence: drozdenko@karlov.mff.cuni.cz

Abstract: Deformation mechanisms in extruded Mg-Y-Zn alloys with different volume fractions of the long-period stacking ordered (LPSO) structure have been investigated during cyclic loading, i.e., compression followed by unloading and reverse tensile loading. Electron backscattered diffraction (EBSD) and in situ neutron diffraction (ND) techniques are used to determine strain path dependence of the deformation mechanisms. The twinning-detwinning mechanism operated in the α-Mg phase is of key importance for the subsequent hardening behavior of alloys with complex microstructures, consisting of α-Mg and LPSO phases. Besides the detailed analysis of the lattice strain development as a function of the applied stress, the dislocation density evolution in particular alloys is determined.

Keywords: Mg-LPSO alloys; neutron diffraction; EBSD; deformation mechanisms; dislocation slip; twinning

Citation: Drozdenko, D.; Farkas, G.; Šimko, P.; Fekete, K.; Čapek, J.; Garcés, G.; Ma, D.; An, K.; Máthis, K. Influence of Volume Fraction of Long-Period Stacking Ordered Structure Phase on the Deformation Processes during Cyclic Deformation of Mg-Y-Zn Alloys. *Crystals* **2021**, *11*, 11. https://dx.doi.org/10.3390/cryst11010011

Received: 30 November 2020
Accepted: 22 December 2020
Published: 25 December 2020

1. Introduction

The development of magnesium (Mg) alloys with long-period stacking ordered (LPSO) structures belong to the most important metallic lightweight alloy innovations of the last two decades. The mechanical properties of such materials surpass those of conventional cast or wrought Mg alloys [1–4].

The LPSO phase-α-Mg matrix interface is reported as a site for dynamic recrystallization, usually leading to the formation of a bimodal grain structure in wrought Mg-Y-Zn alloys [5]. The complex microstructure of extruded Mg-LPSO alloys, including the fine dynamically recrystallized (DRX) and coarse non-DRX grains of the α-Mg phase and the fiber-shaped LPSO-phase, has been found to be beneficial for the resulting mechanical properties [4]. The stiffer LPSO phase reinforces the magnesium matrix, what results in excellent strength values. At the same time, fine DRX grains contribute to the ductility of the alloy.

Naturally, the amount of the alloying elements (Zn and Y) influences the microstructure evolution during both casting and thermo-mechanical processing and determines the volume fraction of the formed LPSO phase, therefore, strongly affecting the mechanical properties [6,7]. Among Mg-LPSO alloys, the best tensile properties at room temperature were achieved for the $Mg_{97}Y_2Zn_1$ (at.%) alloy prepared by powder metallurgy (tensile yield strength of 610 MPa, elongation of 5%) [2]. Using another processing route, such as rapid solidification (RS) of the ribbons and its consolidation via extrusion, the materials having ultrafine grain microstructures with dispersed stacking faults can be produced. The RS-extruded $Mg_{97.94}Zn_{0.56}Y_{1.5}$ (at.%) alloy can exhibit a yield strength of 362 MPa and elongation of 18.2% [8]. It is obvious that with a proper choice of alloy content and processing parameters (e.g., temperature, extrusion ratio, and extrusion rate) the microstructure and mechanical properties of the Mg-LPSO alloys can be varied in a wide range.

Several deformation mechanisms can operate in particular microstructural elements during the straining of wrought Mg-LPSO alloys. Activity of basal slip was reported in DRX α-Mg grains [5]. Concurrently, the non-basal slip as well as the presence of extension $\{10\bar{1}2\}\langle 10\bar{1}\bar{1}\rangle$ twinning has been observed in coarse non-DRX α-Mg grains [5]. Those mechanisms can also be active in DRX α-Mg grains, but at significantly higher applied stresses (the grain size effect) [9]. The LPSO phase is known to be deformed by kinking when concurrently shear and compression deformation act parallel to LPSO fibers. This mechanism is governed by the collective motion of basal dislocations. If the load is applied perpendicular to the LPSO fibers, rather non-basal slip dominates [5]. Thus, besides volume fraction, the orientation and distribution of the LPSO phase significantly affect the mechanical behavior of the material [5,7,10–15]. The orientation effect is given by the texture formed during the processing of wrought Mg-LPSO alloys. In general, both α-Mg and LPSO phases are characterized by their basal planes oriented parallel to extrusion direction (ED) [5]. The preferential activity of deformation mechanisms, with respect to the mutual loading axis and *c*-axis of the lattice, determines the resulting mechanical properties. For example, the polar nature of extension twinning [16–19] leads to extensive twin nucleation during compressive loading along ED (i.e., during compression perpendicular to the *c*-axis of the lattice of the α-Mg phase), and therefore, significantly lower yield stress comparing to one during tensile loading is observed [20]. During subsequent reverse loading of pre-compressed Mg alloy, detwinning could be activated in the twinned fractions [9,21,22]. This process is usually characterized by thickness reduction or complete disappearance of existing twin lamellae as their lattices are rotating back to the original orientation [23]. Thus, if the extension twins occur during in-plane compression along ED, detwinning takes place during subsequent reverse tensile loading. Thus, understanding the cyclic loading behavior is essential for engineering applications. However, there are only a few works dealing with this issue in Mg-LPSO alloys. Hagihara et al. [24] investigated low cycle fatigue properties for several Mg alloys with various content of the LPSO phase. They found that cyclic hardening is not significant. Furthermore, the apparent yield stress (YS) is gradually decreasing with increasing number of cycles for $Mg_{97}Zn_1Y_2$ alloys (in at%) in both (extrusion and transversal) loading directions. In contrast, a gradual increase in YS was found for the tensile side of the cycle in $Mg_{99.2}Zn_{0.2}Y_{0.6}$ alloy. In cast Mg-LPSO alloys, besides the influence of the composition, the effect of compressive pre-loading was studied [25]. Both the compressive pre-loading and the higher LPSO content increased the share of the kinematic hardening to cyclic hardening.

Similar to conventional Mg alloys, in Mg-LPSO-based materials, the twinning-detwinning phenomenon operated in the α-Mg phase plays a significant role during cyclic loading. To investigate this process, besides microstructure observations provided by scanning electron microscopy (SEM) (in situ and *post mortem*) [9,22,26,27], the in situ diffraction method has been found to be a powerful experimental tool [5,27,28]. The twinned volume can be estimated from the intensity variations of particular diffraction peaks [29]. At the same time, the activation stress of various dislocation slip systems in α-Mg and LPSO phases can be deduced from the evaluation of the lattice strain with applied stress [30].

The main aim of the present work is to reveal the single-cycle compression-tension properties of Mg-LPSO alloys with respect to the variation in the volume fraction of the LPSO phase. The experimental approach includes in situ neutron diffraction and mapping of the microstructure by electron backscattered diffraction (EBSD) technique. The novelty of the work consists in the identification of the active deformation mechanisms during the strain path change and revealing the evolution of the dislocation structure in the Mg matrix as a function of the applied load and fraction of the LSPO phase.

2. Experimental

For the present study, three alloy compositions WZ42-$Mg_{98.5}Y_{1.0}Zn_{0.5}$, WZ72-$Mg_{97.0}Y_{2.0}Zn_{1.0}$, and WZ104-$Mg_{95.5}Y_{3.0}Zn_{1.5}$ (nominal compositions in at%) were selected. The master alloys were cast at the Korea Institute of Industrial Technology and further extruded at 350 °C with 0.5 mm s^{-1} extrusion speed and an extrusion ratio of 1:10 in National Center for Metallurgical Research (CENIM-CSIC), Madrid.

The in situ neutron diffraction (ND) measurements were performed on the VULCAN Engineering Materials Diffractometer beamline at Oak Ridge National Laboratory. The cylindrical specimens (thread-ended, gauge length and diameter were 20 and 9 mm, respectively) were fixed horizontally in a deformation rig manufactured by Measure Test Simulate (MTS) company. The specimens were loaded along ED at room temperature. The angle between the incident beam and the specimen was 45°. The diffraction patterns in the axial and radial directions were detected by two stationary detector banks located at ± 90° to the incoming beam. The neutron gauge volume was 245 mm^3. The diffraction patterns were recorded in continuous and discontinuous modes, respectively. In the first case, the deformation cycle was executed without stopping at a strain rate of 10^{-3} s^{-1}, and the diffraction data (patterns) were recorded continuously. For the evaluation, obtained data were binned to 1 min long intervals. Since the proper evaluation of the LPSO peaks and the diffraction line profile analysis of the magnesium peaks requires data with good enough statistics, the tests in the discontinuous mode were stopped at pre-defined strain levels (0.1, 0.5, 1, 2, 3, 4%) for approximately 20 min for the collection of diffraction patterns. The ND patterns recorded in the discontinuous mode were evaluated by the Convolutional Multiple Whole Profile (CMWP) fitting method [31,32]. The diffraction patterns were fitted by functions constructed from the background spline, instrumental pattern, and a theoretical profile function. The theoretical profile function assumes dislocation-caused microstrains. As a result of the fitting procedure, the dislocation density (ρ) has been directly obtained.

The microstructure of the as-extruded material, as well as specimens deformed up to specified strain levels, has been investigated using SEM Quanta FX200 (Field Electron and Ion Company—FEI Company, Hillsboro, OR, USA) and Auriga Compact (Zeiss, Oberkochen, Germany) both equipped with an EBSD camera (EDAX Inc., Mahwah, NJ, USA). The specimen for microscopy observations were ground on SiC papers and subsequently polished by diamond paste with a particle size decreasing down to 0.25 µm. The final step of specimen preparation consisted of ion polishing using a Leica EM RES102 device.

3. Results and Discussion

The initial microstructure and texture of the specimens are presented in Figure 1. Investigated alloys are characterized by microstructure consisting of the LPSO phase (light contrast in backscatter electron (BSE) images) and α-Mg grains (dark contrast in BSE images), including small DRX and coarse non-DRX grains elongated along ED. The volume fractions of the LPSO phase has been estimated from the BSE images as 10, 21, and 35% for the WZ42, WZ72, and WZ104 alloys, respectively. The fraction of coarse non-DRX grains and their grain size decreases with the increasing amount of the alloying elements from the WZ42 alloy towards the WZ104 alloy. Those elongated grains in all of the studied alloys have an intensive texture with their *c*-axis perpendicular to ED. (Those few grains represented in the EBSD maps contributes to the higher texture intensities along the periphery of (0001) pole figures). Although the DRX grains show a more random orientation distribution, the overall texture has an ordered character with the basal planes oriented parallel to ED, see pole figures composed from EBSD maps and rather more informative intensity distribution evaluated from ND data in Figure 1d. The decrease in texture intensity with increasing alloying elements can be associated with the decrease in the volume fraction of coarse non-DRX α-Mg grains.

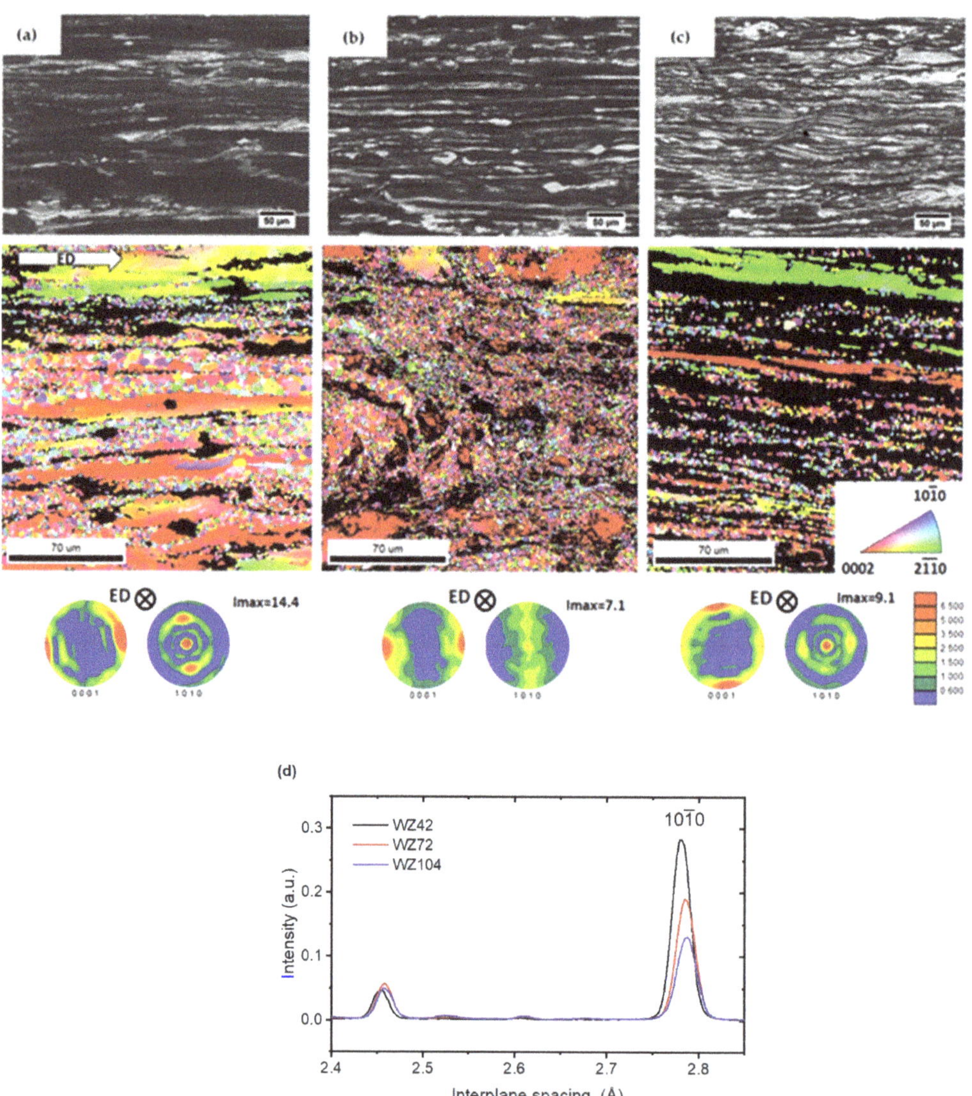

Figure 1. Initial microstructures and texture of the (**a**) WZ42, (**b**) WZ72, (**c**) WZ104 alloys, the intensity of the $\{10\bar{1}0\}$ peak evaluated from ND data (**d**).

The obtained EBSD maps are indexed for α-Mg phase and black areas in the maps (related to pixels with a confidence index for α-Mg < 0.1) are therefore associated with the LPSO phase. The volume fractions of the LPSO phase estimated by BSE images are in good agreement with those obtained from the EBSD maps.

The deformation curves are presented in Figure 2. (The small waves on the curves around 0 MPa during unloading are given by the backlash of the deformation setup.) It is obvious that the flow stress increases with increasing volume fraction of the LPSO phase. First, during the compression part, a plateau follows the macroscopic yield, especially pronounced for WZ42 (see the inserted part in Figure 2), which can be a sign of enhanced activity of $\{10\bar{1}2\}\langle10\bar{1}\bar{1}\rangle$ extension twinning [20]. During reverse tension, the stress–strain curve of the WZ42 specimen shows a pronounced S-shape, indicating a significant role of detwinning [9,33]. In the case of the WZ72 and WZ104 alloys, the S-shape is not observed, and their stress-strain curves have convex shape, typically observed during tensile loading of Mg alloys.

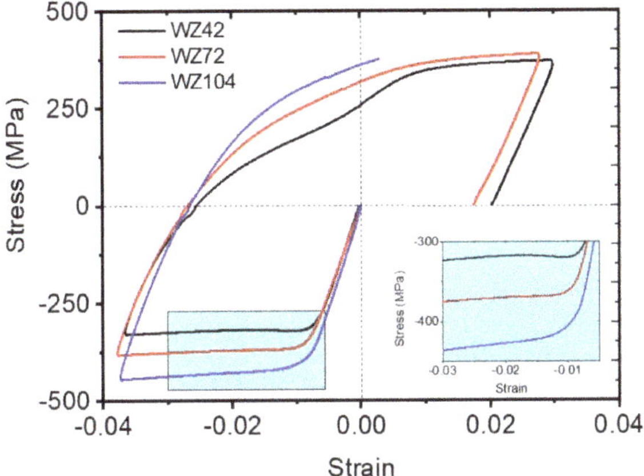

Figure 2. The deformation curves for the one-cycle deformation of the Mg-LPSO alloys.

The microstructures of the alloys after the full-cycle deformation are presented in Figure 3. In the WZ42 alloy, cracks in the LPSO lamellae formed during straining are observed, Figure 3a. Kinking is characteristic for the WZ72 and WZ104 alloys, Figure 3b,c. The result of high twinning activity in WZ42 alloy can be spotted on the image quality (IQ) maps in Figure 3d. There is a relatively high number of leftover twin boundaries (highlighted by the red color). The twin boundaries are identified as extension $\{10\bar{1}2\}\langle10\bar{1}\bar{1}\rangle$ twins with a misorientation angle of 86.3° with respect to the original lattice. In the case of the WZ72 alloy, this microstructural feature is negligible (cf. Figure 3e), and no twin boundaries have been observed in the WZ104 alloy (not presented here), which can be explained by the decreased amount and grain size of α-Mg grains in this alloy.

The appropriate volume fraction and grain size of non-DRX grains in the WZ42 alloy give a rise to investigate the evolution of twinning at particular stages of cyclic loading using the EBSD/SEM technique with high confidence. Therefore, an additional specimen was deformed by the same deformation loading cycle like those for ND tests. The microstructure has been examined just above the macroscopic yield, at 330 MPa of compression stress, and after a reverse tension up to 225 MPa, Figure 4. It can be seen that, after reaching the yield point, the twins appear mainly in elongated non-DRX grains (Figure 4a) and several twin variants are nucleated within a particular grain, forming families of twins (groups of twins with same orientation with respect to original grain). A schematic view in Figure 4a represents activated variants of twins with respect to the orientation of the parent grain. It can be seen that the tilt of nucleated twin lamellae with respect to ED (lenticular twin lamella is perpendicular to ED in "red-colored" grains and tilted by 45° from ED in "green-yellow" grains) is given by orientation of the twin plane with respect to the observation point of view. With increasing load (Figure 4b), the already existing twins become thicker, and new thin twins occur as well. It should be noted that, even at the compression peak stress, the non-DRX grains are still not fully twinned. Thus, twin boundaries are supposed to be mobile for further forward or backward movement. At the same time, twins in the small DRX grains can be also observed at higher compressive stresses, cf. BSE image in Figure 4b. During the reverse-tensile loading (Figure 4c), the twins nucleated during compression become thinner, and only a few narrow twins can be found in some non-DRX grains. The leftovers of these twin boundaries can be seen in Figure 4c. Besides, the trace of twin boundaries after complete detwinning are also seen, see gray lines in the white rectangular in Figure 4c. This could be explained by the localization of strain inside the lattice due to a high concentration of dislocations, which have been acting for twin growth and shrinkage. The operation of twins is also well seen in the texture development (see pole figures in Figure 4). In the initial state, basal planes are oriented parallel to ED, therefore texture intensity is distributed at the periphery of the (0001) pole figure. With increasing applied stress, a strong texture component in the middle of (0001) pole figure is formed due to the rotation of basal planes by almost 90° from the original orientation toward ED as a result of $\{10\bar{1}2\}\langle10\bar{1}\bar{1}\rangle$ twinning. The intensity of this texture component increases with increasing compressive loading, what can be associated with massive twin growth. The texture of the specimen after reverse-tensile loading is comparable to the one in the initial state. The disappearance of the texture component is given by detwinning, i.e., rotation of the lattice in the twin fractions back to the original orientation.

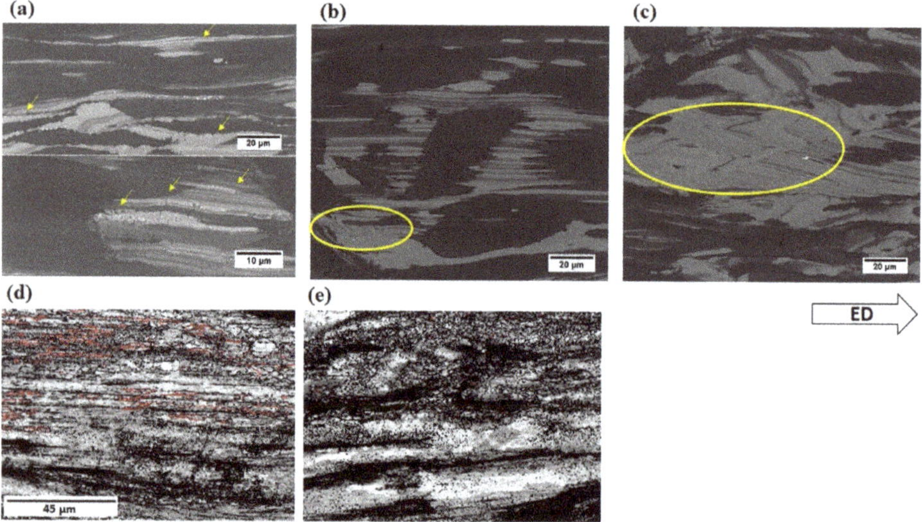

Figure 3. The microstructures of (**a**,**d**) WZ42, (**b**,**e**) WZ72, and (**c**) WZ104 alloys after full-cycle deformation. Kinking and cracks are highlighted by yellow marks in backscatter electron (BSE) images (**a**–**c**). Boundaries of tensile twins are marked in red color in image quality (IQ) maps in (**d**,**e**).

Owing to the high fraction of the LPSO phase and a significantly low fraction of non-DRX grains, similar EBSD/SEM analysis is rather difficult for the WZ72 and WZ104 specimens. Thus, from a statistical point of view, the ND data, which characterize a large specimen volume, gives a better insight into the twinning development.

The change of the intensity of the (0002)–$\{10\bar{1}0\}$ diffraction peak pair is directly related to the extension twinning [29]. With respect to the initial texture of specimens and the diffraction geometry used in our experiments, the intensity of the (0002) peak is expected to increase in the axial detector, whereas the intensity of the $\{10\bar{1}0\}$ peak should decrease once the extension twinning is active. In Figure 5, the intensity changes in the axial direction of (0002)–$\{10\bar{1}0\}$ peaks are plotted as a function of the applied stress and strain. It can be seen that, in compression, the twinned volume starts to increase at significantly lower stress in WZ42 than that in the WZ72 and WZ104 specimens. Moreover, there is an effect of volume fraction of the LPSO phase (i.e., alloy content) on the resulting twinned volume; the twin volume fraction for WZ42 is significantly larger compared to that for the other two alloys. During the unloading, detwinning took place, however, it is not finished at zero stress and continue during tension. The intensity changes plotted against strain unambiguously indicate that the detwinning is terminated after reaching the zero-strain value (cf. Figure 5c).

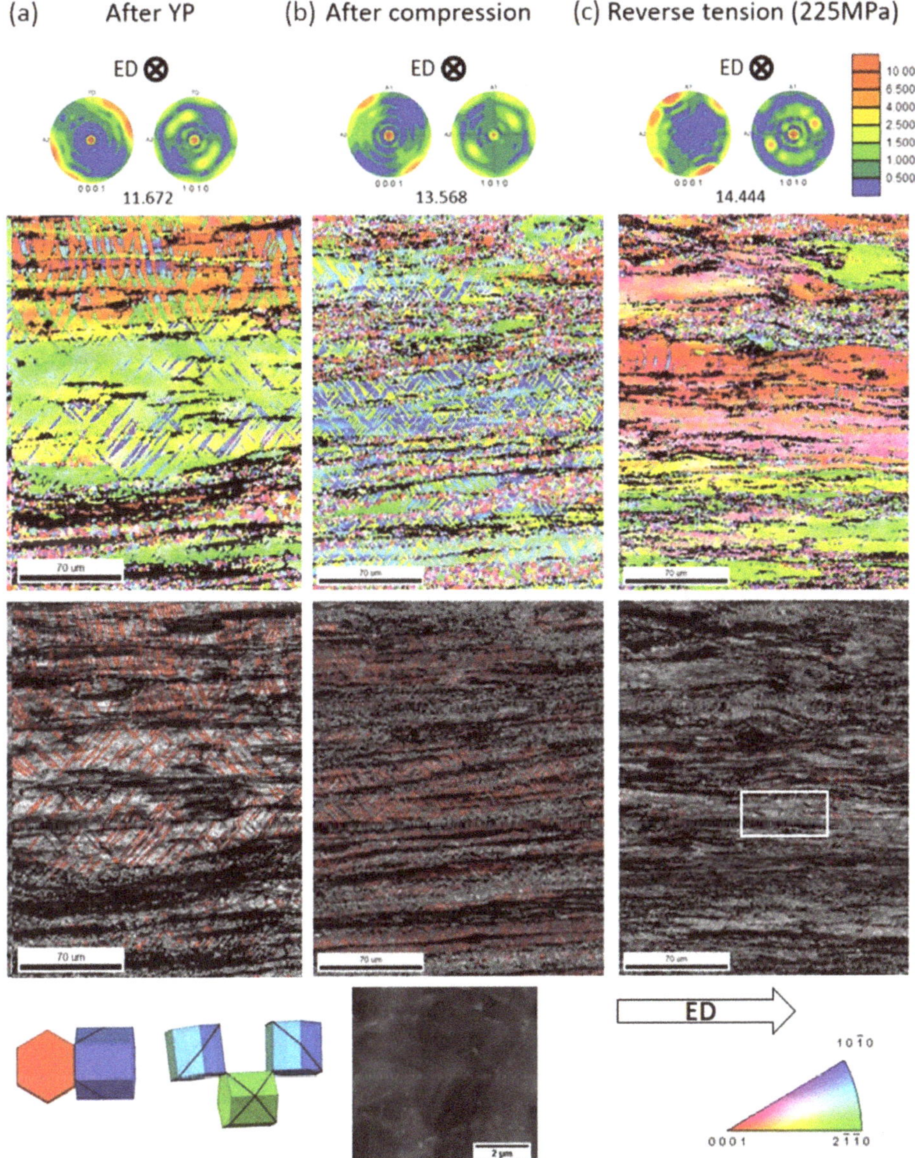

Figure 4. Development of the microstructure during cyclic loading of WZ42 alloy, particularly after (**a**) yield point (YP), (**b**) compression part, and (**c**) reverse tensile loading up to 225 MPa. Schematic view of hexagonal close-packed lattices are used for representation of activated twin variants with respect to parent grain orientation. Extra BSE image represents activation of twinning in dynamically recrystallized (DRX) grains.

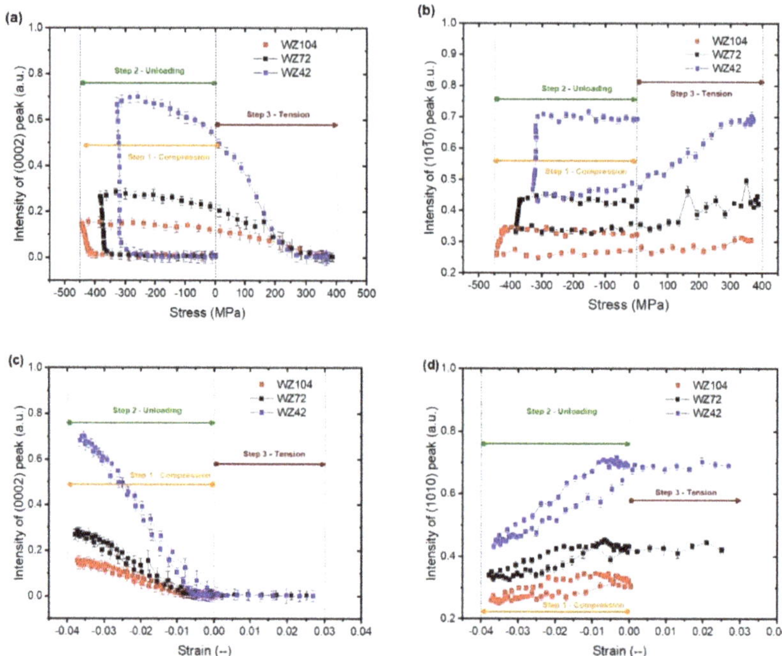

Figure 5. The intensity changes of the (0002)–$\{10\bar{1}0\}$ peaks as a function of applied stress (**a**,**b**) and strain (**c**,**d**). Data from axial detector.

The lattice strains plotted for axial detector as a function of applied stress, Figure 6, are determined using the equation

$$\varepsilon = \frac{d - d_0}{d_0}, \tag{1}$$

where d, d_0 are the lattice spacing for deformed and stress-free conditions, respectively. In this representation, a deviation from the linear Hooke's elasticity indicates activation of specific deformation mechanisms. However, due to the plateau stress present for all alloys above the yield point, the applied stress-lattice strain plot does not provide an easy survey. Therefore, the applied strain-lattice strain plots are also shown here, which are more representative in the region of the plastic deformation.

The stress evolution of the (0002)–$\{10\bar{1}0\}$ lattice strain indicates the activity of extension twinning with respect to the composition of the alloys, and therefore the volume fraction of the LPSO phase. The lattice strain evolution is in a good agreement with the development of intensity of the (0002)–$\{10\bar{1}0\}$ diffraction peak pair presented in Figure 5.

The onset of the twinning takes place far below the macroscopic strain for the WZ42 and WZ72 specimens, and the activation stress increases with the increasing amount of alloying elements. The (0002) grains (that is, grains with their (0002) axis along the axial direction) are in "soft-orientation" during compression, and they relax after twin initiation, whereas the "hard-oriented" $\{10\bar{1}0\}$ grains accommodate a higher portion of elastic loading. The relaxation is well seen in the applied strain-lattice strain plot (Figure 6d,e).

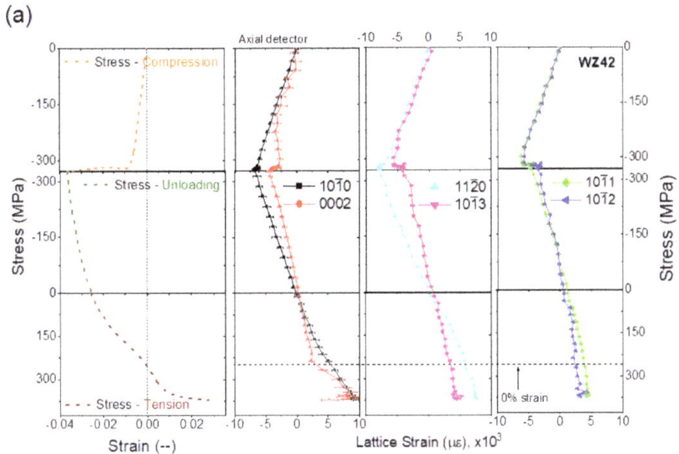

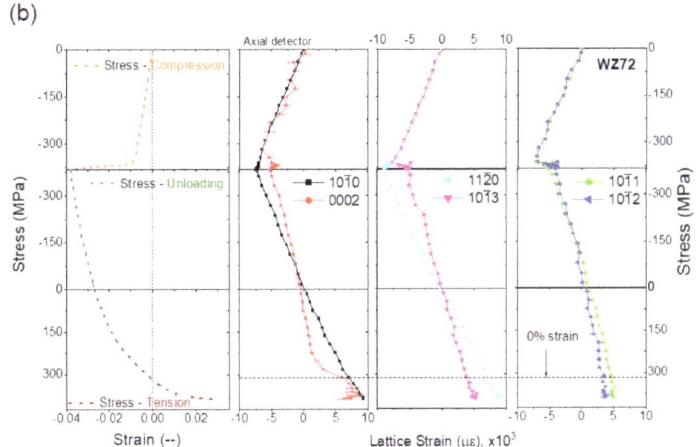

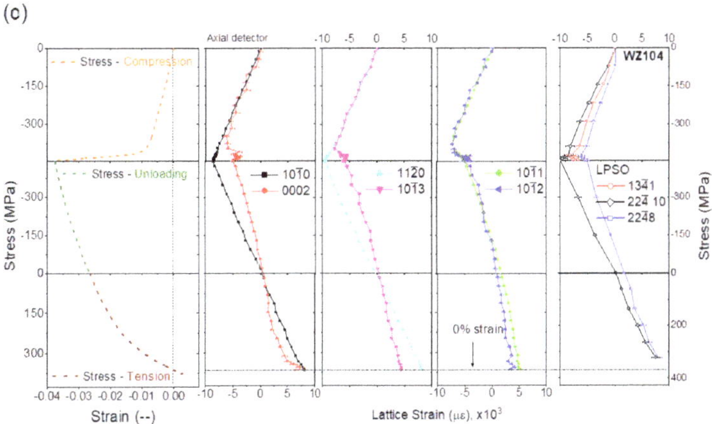

Figure 6. *Cont.*

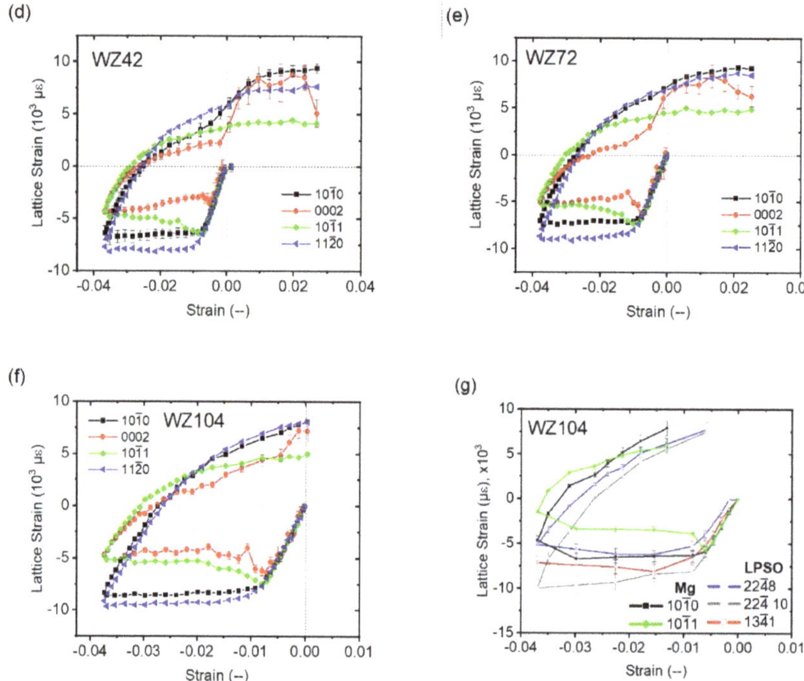

Figure 6. The elastic lattice strains plotted for axial detection as a function of applied stress and strain for the (**a**,**d**) WZ42, (**b**,**e**) WZ72, and (**c**,**f**,**g**) WZ104 alloys. The lattice strains for the long-period stacking ordered (LPSO) peaks were evaluated from the discontinuous measurement data.

During unloading, the compressive lattice strains in these grains relax. In the tension part, the soft–hard orientation roles interchange owing to the polar nature of the extension twinning [16–19]. Consequently, a hardening of (0002) oriented grains is observed, which are now unfavorably oriented for extension twinning with respect to the loading direction. This feature is the most significant for the WZ42 alloy. For the $\{11\bar{2}0\}$ orientation, the contribution of prismatic and pyramidal <*a*>-slip also cannot be excluded (Schmid-factor values for the slip in these systems are 0.43 and 0.38, respectively). For the WZ42 specimen in tension, the $\{11\bar{2}0\}$ lattice strain relaxes around 150 MPa, indicating significant non-basal <*a*>-slip.

In the case of the WZ104 specimen, the $\{4\bar{2}\bar{2}8\}$ $\{4\bar{2}\bar{2}10\}$, and $\{41\bar{3}1\}$ LSPO diffraction peaks were intensive enough for confident lattice strain evaluation. As it can be seen from Figure 6c,f,g, the $\{4\bar{2}\bar{2}8\}$ and $\{41\bar{3}1\}$ planes of the LPSO phase shares the highest load compared to other ($\{4\bar{2}\bar{2}10\}$) LPSO- and α-Mg-related planes below the yield point. In the plastic region, around -0.015 of applied strain further softening of the $\{4\bar{2}\bar{2}10\}$ planes takes place, whereas hardening of $\{41\bar{3}1\}$ is observed. In [34], it was shown that the change in the lattice strain distribution on the $\{4\bar{2}\bar{2}10\}$ planes can be associated with deformation by kinking. During unloading and reverse-tensile loading, relaxation takes place.

The evolution of the dislocation density in the α-Mg phase as a function of the applied stress for the investigated alloys is plotted in Figure 7. In the compression part, all of them behave similarly; above the yield point, the dislocation density increases due to continuously active dislocation slip systems. The highest ρ value is reached for the WZ104 alloy. After change of strain path (direction), for all investigated alloys, ρ decreases, what can be associated with annihilation of dislocation due the relaxation of internal stresses and closing sources of dislocations activated during compressive load. However, after reaching a certain stress level, the decrease in ρ terminates thanks to the activation of new sources of dislocations activated as a result of tensile loading. In the WZ42 alloy, detwinning significantly affects dislocation density development, Figure 5a. In the grain fraction, which reorients to the initial orientation, further slip is possible. Therefore, detwinning during tension is accompanied by a slight increment of the dislocation density. However, complete detwinning causes annihilation of a high number of dislocations on twin boundaries, leading to decrease in dislocation density in the secondary hardening part.

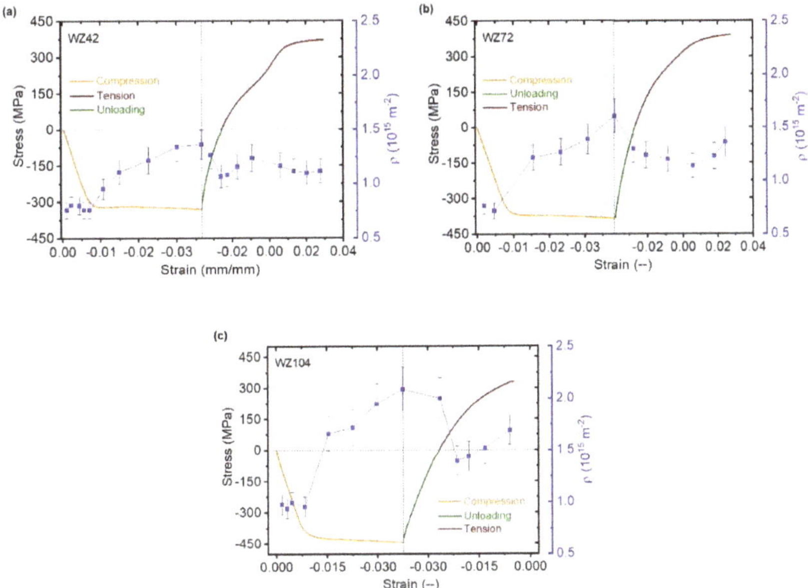

Figure 7. The evolution of the dislocation density in α-Mg phase as a function of applied stress for the (**a**) WZ42, (**b**) WZ72, and (**c**) WZ104 alloys.

The results listed above indicate that the volume fraction of the LPSO phase significantly influences the deformation behavior of the Mg-LPSO alloys. For the WZ42 specimen, where the fraction of the LPSO phase is the lowest, the bimodal grain structure enhances the extension twinning. According to the observations for wrought Mg alloys [35,36], the coarse non-DRX grains twin first. The reason for this behavior is given by the low internal stress in those grains and by the long grain boundary, which serves as a nucleation site for twinning [37]. In contrast, the WZ104 alloy is characterized by a 35% fraction of the LPSO phase and a significantly smaller fraction of non-DRX grains than that in the WZ42 alloy (Figure 1). Accordingly, the highest twinned volume can be observed in the WZ42 alloy. During the unloading, detwinning takes place in all investigated alloys, especially pronounced for the WZ42 alloy. The detwinning is enhanced due to the limited twin thickening during compressive load as a result of the stored internal stress introduced into the alloy by the presence of the LPSO phase. Therefore, the majority of twins are in unrelaxed conditions, leading to their easier disappearance during reverse loading [9,38]. A detailed inspection of the lattice strain evolution for WZ42 and WZ72 specimen suggests that, in compression, the governing mechanism at the onset of plasticity is extension twinning (with significant dominance in WZ42). Basal slip also takes place, but its share is smaller than that of twinning. This finding is in good agreement with the work of Vinogradov et al. [39] in wrought ZK60 magnesium alloy, where dominance of twinning at the beginning of the plastic flow has been revealed by analysis of the acoustic emission signal. However, the macroscopic yield needs the activation of the non-basal <*a*>-slip as well, leading to the rapid increment of the dislocation density (cf. Figure 7) and relaxation of the lattice strains on the $\{10\bar{1}1\}$ and $\{10\bar{1}2\}$ grain families. At the beginning of reverse tension, the detwinning is dominating. Nevertheless, the <a>-slip on both the basal and non-basal planes is significant. Owing to the basal texture, significantly fewer grains are favorably oriented for extension twinning in tension than in compression. Consequently, the role of dislocation slip in plasticity increases.

In the WZ104 alloy, the development of the deformation mechanisms is different. Owing to the high fraction of the LSPO phase, a composite-like behavior is observed when the LSPO phase shares a larger portion of the load than the magnesium matrix. This behavior can be followed in Figure 6c. Due to this stress shielding effect, the twin nucleation, which is also suppressed by the small grain size of α-Mg grains and large internal stresses, can start only around the yield point. In compression, above the yield point, the lattice strain in the LPSO phase relaxes, indicating the deformation of this phase by the kinking mechanism, Figure 3c. During reverse tension, it is obvious (see Figure 6c) that the role of extension twinning is negligible. The strain is accommodated by <*a*>-slip and also by deformation of the LPSO phase. However, in tension, the kink formation is unlikely, and the LPSO phase is deformed rather by non-basal <a>-slip [40,41]. The enhanced dislocation activity both in the LPSO-phase and the magnesium matrix leads to high internal stresses and early failure of the material.

4. Conclusions

The in situ neutron diffraction (ND) technique has been employed to determine strain path dependences of the deformation mechanisms in Mg-Y-Zn alloys with various volume fractions of the LPSO phase.

The obtained ND data gives insight into the twinning-detwinning mechanisms operating in the investigated alloys. This mechanism is mainly realized in the non-DRX α-Mg grains and affects the overall development of deformation behavior of the Mg-LPSO alloys. It is the most significant in the WZ42 alloy with the high-volume fraction of α-Mg grains at the expense of the lowest volume fraction of the LPSO phase. The WZ104 alloy, having a high-volume fraction of the LPSO phase, behaves as a composite material, and the LPSO phase bears the main part of the applied compressive load. Further, the non-basal slip and kinking dominate the plastic deformation. Consequently, the twinning–detwinning mechanism plays a minor role, and the strain in the α-Mg-matrix is rather accommodated by non-basal slip.

Additionally, the analysis of ND profiles provides information about the development of dislocation density during cyclic loading. After reaching the yield point in compression, the dislocation density increases in all alloys, followed by a decrease during unloading. The highest dislocation density develops in the WZ104 alloy owing to the low twinning activity.

Author Contributions: K.M., G.G., and J.Č. conceived and designed the experiments; J.Č., D.M., K.A., and K.M. performed the neutron diffraction tests; D.D. and K.F., studied the microstructure using SEM/EBSD; D.D., G.F., P.Š., K.F., J.Č., and K.M. performed formal analysis and contributed to discussion; K.M. and D.D. worked on writing—original draft preparation; G.F., K.F., J.Č., G.G., D.M., and K.A. revised the manuscript. All authors have read and agreed to the published version of the manuscript.

Funding: This research was funded by the Czech Science Foundation under grant 20-07384Y (D.D., G.F., K.F.); by Spanish Ministry of Economy and Competitiveness under project number MAT2016-78850-R; and by the Operational Programme Research, Development and Education, The Ministry of Education, Youth and Sports (OP RDE, MEYS) under the grant CZ.02.1.01/0.0/0.0/16_013/0001794.

Institutional Review Board Statement: Not applicable.

Informed Consent Statement: Not applicable.

Data Availability Statement: The data that supports the findings of the study is available from the corresponding author, D.D., upon reasonable request.

Acknowledgments: We acknowledge Oak Ridge National Laboratory (Oak Ridge, TN, USA) for the provision of experimental facility. Parts of this research were carried out at ORNL under the proposal number 16584.

Conflicts of Interest: The authors declare no conflict of interest.

References

1. Inoue, A.; Kawamura, Y.; Matsushita, M.; Hayashi, K.; Koike, J. Novel hexagonal structure and ultrahigh strength of magnesium solid solution in the Mg–Zn–Y system. *J. Mater. Res.* **2001**, *16*, 1894–1900. [CrossRef]
2. Kawamura, Y.; Hayashi, K.; Inoue, A.; Masumoto, T. Rapidly Solidified Powder Metallurgy $Mg_{97}Zn_1Y_2$ Alloys with Excellent Tensile Yield Strength above 600 MPa. *Mater. Trans.* **2001**, *42*, 1172–1176. [CrossRef]
3. Kawamura, Y.; Kasahara, T.; Izumi, S.; Yamasaki, M. Elevated temperature Mg97Y2Cu1 alloy with long period ordered structure. *Scr. Mater.* **2006**, *55*, 453–456. [CrossRef]
4. Yamasaki, M.; Hashimoto, K.; Hagihara, K.; Kawamura, Y. Effect of multimodal microstructure evolution on mechanical properties of Mg–Zn–Y extruded alloy. *Acta Mater.* **2011**, *59*, 3646–3658. [CrossRef]
5. Garcés, G.; Máthis, K.; Medina, J.; Horváth, K.; Drozdenko, D.; Oñorbe, E.; Dobroň, P.; Pérez, P.; Klaus, M.; Adeva, P. Combination of in-situ diffraction experiments and acoustic emission testing to understand the compression behavior of Mg-Y-Zn alloys containing LPSO phase under different loading conditions. *Int. J. Plast.* **2018**, *106*, 107–128. [CrossRef]
6. Horvath, K.; Drozdenko, D.; Garces, G.; Mathis, K.; Dobron, P. Effect of Extrusion Ratio on Microstructure and Resulting Mechanical Properties of Mg Alloys with LPSO Phase. In *Magnesium Technology 2017*; Springer: Berlin/Heidelberg, Germany, 2017; pp. 29–34. [CrossRef]
7. Horváth, K.; Drozdenko, D.; Daniš, S.; Garcés, G.; Máthis, K.; Kim, S.; Dobroň, P. Characterization of Microstructure and Mechanical Properties of Mg–Y–Zn Alloys with Respect to Different Content of LPSO Phase. *Adv. Eng. Mater.* **2018**, *20*, 1700396. [CrossRef]
8. Drozdenko, D.; Yamasaki, M.; Máthis, K.; Dobroň, P.; Lukáč, P.; Kizu, N.; Inoue, S.-I.; Kawamura, Y. Optimization of mechanical properties of dilute Mg-Zn-Y alloys prepared by rapid solidification. *Mater. Des.* **2019**, *181*, 107984. [CrossRef]
9. Drozdenko, D.; Bohlen, J.; Yi, S.; Minárik, P.; Chmelík, F.; Dobroň, P. Investigating a twinning–detwinning process in wrought Mg alloys by the acoustic emission technique. *Acta Mater.* **2016**, *110*, 103–113. [CrossRef]
10. Garces, G.; Perez, P.; Cabeza, S.; Lin, H.K.; Kim, S.; Gan, W.; Adeva, P. Reverse tension/compression asymmetry of a Mg–Y–Zn alloys containing LPSO phases. *Mater. Sci. Eng. A* **2015**, *647*, 287–293. [CrossRef]
11. Garces, G.; Muñoz-Morris, M.A.; Morris, D.G.; Jimenez, J.A.; Perez, P.; Adeva, P. The role of extrusion texture on strength and its anisotropy in a Mg-base alloy composed of the Long-Period-Structural-Order phase. *Intermetallics* **2014**, *55*, 167–176. [CrossRef]
12. Hagihara, K.; Kinoshita, A.; Sugino, Y.; Yamasaki, M.; Kawamura, Y.; Yasuda, H.Y.; Umakoshi, Y. Effect of long-period stacking ordered phase on mechanical properties of Mg97Zn1Y2 extruded alloy. *Acta Mater.* **2010**, *58*, 6282–6293. [CrossRef]
13. Horváth, K.; Drozdenko, D.; Garcés, G.; Dobroň, P.; Máthis, K. Characterization of the acoustic emission response and mechanical properties of Mg alloy with LPSO phase. *Mater. Sci. Forum* **2017**, *879*, 762–766. [CrossRef]
14. Hagihara, K.; Yokotani, N.; Umakoshi, Y. Plastic deformation behavior of Mg12YZn with 18R long-period stacking ordered structure. *Intermetallics* **2010**, *18*, 267–276. [CrossRef]
15. Oñorbe, E.; Garcés, G.; Pérez, P.; Adeva, P. Effect of the LPSO volume fraction on the microstructure and mechanical properties of Mg–Y2X–Zn X alloys. *J. Mater. Sci.* **2012**, *47*, 1085–1093. [CrossRef]
16. Partridge, P.G. The crystallography and deformation modes of hexagonal close-packed metals. *Metall. Rev.* **1967**, *12*, 169–194. [CrossRef]
17. Barnett, M.R. Twinning and the ductility of magnesium alloys Part I: "Tension" twins. *Mater. Sci. Eng. A* **2007**, *464*, 1–7. [CrossRef]
18. Kelley, E.W.; Hosford, W.F. Plane-strain compression of magnesium and magnesium alloy crystals. *Trans. AIME* **1968**, *242*, 5–13.
19. Brown, D.W.; Agnew, S.R.; Bourke, M.A.M.; Holden, T.M.; Vogel, S.C.; Tome, C.N. Internal strain and texture evolution during deformation twinning in magnesium. *Mater. Sci. Eng. A* **2005**, *399*, 1–12. [CrossRef]
20. Kelley, E.W.; Hosford, W.F. The deformation characteristics of textured magnesium. *Trans. AIME* **1968**, *242*, 654–661.
21. Wang, Y.N.; Huang, J.C. The role of twinning and untwinning in yielding behavior in hot-extruded Mg-Al-Zn alloy. *Acta Mater.* **2007**, *55*, 897–905. [CrossRef]
22. Lou, X.Y.; Li, M.; Boger, R.K.; Agnew, S.R.; Wagoner, R.H. Hardening evolution of AZ31B Mg sheet. *Int. J. Plast.* **2007**, *23*, 44–86. [CrossRef]
23. Chapuis, A.; Xin, Y.C.; Zhou, X.J.; Liu, Q. {10–12} Twin variants selection mechanisms during twinning, re-twinning and detwinning. *Mater. Sci. Eng. A* **2014**, *612*, 431–439. [CrossRef]
24. Hagihara, K.; Kinoshita, A.; Sugino, Y.; Yamasaki, M.; Kawamura, Y.; Yasuda, H.Y.; Umakoshi, Y. Plastic deformation behavior of Mg97Zn1Y2 extruded alloys. *Trans. Nonferrous Met. Soc. China* **2010**, *20*, 1259–1268. [CrossRef]
25. Shiraishi, K.; Mayama, T.; Yamasaki, M.; Kawamura, Y. Strain-hardening behavior and microstructure development in polycrystalline as-cast Mg-Zn-Y alloys with LPSO phase subjected to cyclic loading. *Mater. Sci. Eng. A* **2016**, *672*, 49–58. [CrossRef]
26. Drozdenko, D.; Čapek, J.; Clausen, B.; Vinogradov, A.; Máthis, K. Influence of the solute concentration on the anelasticity in Mg-Al alloys: A multiple-approach study. *J. Alloys Compd.* **2019**, *786*, 779–790. [CrossRef]

27. Fekete, K.H.; Drozdenko, D.; Čapek, J.; Máthis, K.; Tolnai, D.; Stark, A.; Garcés, G.; Dobroň, P. Hot deformation of Mg-Y-Zn alloy with a low content of the LPSO phase studied by in-situ synchrotron radiation diffraction. *J. Magnes. Alloy.* **2020**, *8*, 199–209. [CrossRef]
28. Muransky, O.; Sittner, P.; Zrnik, J.; Oliver, E.C. In situ neutron diffraction investigation of the collaborative deformation-transformation mechanism in TRIP-assisted steels at room and elevated temperatures. *Acta Mater.* **2008**, *56*, 3367–3379. [CrossRef]
29. Gharghouri, M.A.; Weatherly, G.C.; Embury, J.D.; Root, J. Study of the mechanical properties of Mg-7.7at.% Al by in-situ neutron diffraction. *Philos. Mag. A* **1999**, *79*, 1671–1695. [CrossRef]
30. Agnew, S.R.; Brown, D.W.; Tome, C.N. Validating a polycrystal model for the elastoplastic response of magnesium alloy AZ31 using in situ neutron diffraction. *Acta Mater.* **2006**, *54*, 4841–4852. [CrossRef]
31. Ribárik, G.; Ungár, T.; Gubicza, J.; Ârik, G.; Unga, T. MWP-fit: A Program for Multiple Whole-Profile Fitting of Diffraction Peak Profiles by ab Initio Theoretical Functions. *J. Appl. Crystallogr.* **2001**, *34*, 669–676. [CrossRef]
32. Balogh, L.; Tichy, G.; Ungár, T. Twinning on pyramidal planes in hexagonal close packed crystals determined along with other defects by X-ray line profile analysis. *J. Appl. Crystallogr.* **2009**, *42*, 580–591. [CrossRef]
33. Bohlen, J.; Dobron, P.; Nascimento, L.; Parfenenko, K.; Chmelik, F.; Letzig, D. The Effect of Reversed Loading Conditions on the Mechanical Behaviour of Extruded Magnesium Alloy AZ31. *Acta Phys. Pol. A* **2012**, *122*, 444–449. [CrossRef]
34. Fekete, K.; Farkas, G.; Drozdenko, D.; Tolnai, D.; Strark, A.; Dobroň, P.; Garcés, G.; Máthis, K. The temperature effect on the plastic deformation of the Mg88Zn7Y5 alloy with LPSO phase studied by in-situ synchrotron radiation diffraction. *Intermetallics*, 2020; under review.
35. Dobron, P.; Chmelik, F.; Yi, S.B.; Parfenenko, K.; Letzig, D.; Bohlen, J. Grain size effects on deformation twinning in an extruded magnesium alloy tested in compression. *Scr. Mater.* **2011**, *65*, 424–427. [CrossRef]
36. Bohlen, J.; Dobron, P.; Swiostek, J.; Letzig, D.; Chmelik, F.; Lukac, P.; Kainer, K.U. On the influence of the grain size and solute content on the AE response of magnesium alloys tested in tension and compression. *Mater. Sci. Eng. A* **2007**, *462*, 302–306. [CrossRef]
37. Beyerlein, I.J.; Capolungo, L.; Marshall, P.E.; McCabe, R.J.; Tome, C.N. Statistical analyses of deformation twinning in magnesium. *Philos. Mag.* **2010**, *90*, 2161–2190. [CrossRef]
38. Siska, F.; Stratil, L.; Cizek, J.; Ghaderi, A.; Barnett, M. Numerical analysis of twin thickening process in magnesium alloys. *Acta Mater.* **2017**, *124*, 9–16. [CrossRef]
39. Vinogradov, A.; Orlov, D.; Danyuk, A.; Estrin, Y. Effect of grain size on the mechanisms of plastic deformation in wrought Mg-Zn-Zr alloy revealed by acoustic emission measurements. *Acta Mater.* **2013**, *61*, 2044–2056. [CrossRef]
40. Matsumoto, R.; Uranagase, M.; Miyazaki, N. Molecular Dynamics Analyses of Deformation Behavior of Long-Period-Stacking-Ordered Structures. *Mater. Trans.* **2013**, *54*, 686–692. [CrossRef]
41. Matsumoto, R.; Uranagase, M. Deformation Analysis of the Long-Period Stacking-Ordered Phase by Using Molecular Dynamics Simulations: Kink Deformation under Compression and Kink Boundary Migration under Tensile Strain. *Mater. Trans.* **2015**, *56*, 957–962. [CrossRef]

Article

Restoration Mechanisms at Moderate Temperatures for As-Cast ZK40 Magnesium Alloys Modified with Individual Ca and Gd Additions

Ricardo Henrique Buzolin [1,2,*], Leandro Henrique Moreno Guimaraes [3], Julián Arnaldo Ávila Díaz [4], Erenilton Pereira da Silva [5], Domonkos Tolnai [6], Chamini L. Mendis [7], Norbert Hort [6] and Haroldo Cavalcanti Pinto [3]

1 Christian Doppler Laboratory for Design of High-Performance Alloys by Thermomechanical Processing, 8010 Graz, Austria
2 Institute of Materials Science, Joining and Forming, Graz University of Technology, 8010 Graz, Austria
3 Department of Materials Engineering, University of São Paulo, São Carlos 13563-120, Brazil; leandrohmg@hotmail.com (L.H.M.G.); haroldo@sc.usp.br (H.C.P.)
4 Campus of São João da Boa Vista, São Paulo State University, São João da Boa Vista 13876-750, Brazil; julian.avila@unesp.br
5 Institute of Engineering, Science and Technology (IECT), Federal University of Vales do Jequitinhonha e Mucuri (UFVJM), Janaúba 39440-000, Brazil; erenilton.silva@ufvjm.edu.br
6 Magnesium Innovation Centre, Helmholtz-Zentrum Geesthacht, D 21502 Geesthacht, Germany; domonkos.tolnai@hzg.de (D.T.); norbert.hort@hzg.de (N.H.)
7 Brunel Centre for Advanced Solidification Technology, Brunel University London, Middlesex UB8 3PH, UK; chamini.mendis@brunel.ac.uk
* Correspondence: ricardo.buzolin@tugraz.at; Tel.: +43-316-873-1612

Received: 28 November 2020; Accepted: 15 December 2020; Published: 16 December 2020

Abstract: The deformation behaviour of as-cast ZK40 alloys modified with individual additions of Ca and Gd is investigated at 250 °C and 300 °C. Compression tests were carried out at 0.0001 s^{-1} and 0.001 s^{-1} using a modified Gleeble system during in-situ synchrotron radiation diffraction experiments. The deformation mechanisms are corroborated by post-mortem investigations using scanning electron microscopy combined with electron backscattered diffraction measurements. The restoration mechanisms in α-Mg are listed as follows: the formation of misorientation spread within α-Mg, the formation of low angle grain boundaries via dynamic recovery, twinning, as well as dynamic recrystallisation. The Gd and Ca additions increase the flow stress of the ZK40, which is more evident at 0.001 s^{-1} and 300 °C. Dynamic recovery is the predominant restoration mechanism in all alloys. Continuous dynamic recrystallisation only occurs in the ZK40 at 250 °C, competing with discontinuous dynamic recrystallisation. Discontinuous dynamic recrystallisation occurs for the ZK40 and ZK40-Gd. The Ca addition hinders discontinuous dynamic recrystallisation for the investigated temperatures and up to the local achieved strain. Gd addition forms a semi-continuous network of intermetallic compounds along the grain boundaries that withstand the load until their fragmentation, retarding discontinuous dynamic recrystallisation.

Keywords: magnesium alloys; deformation behaviour; restoration mechanisms; electron microscopy; characterisation; in-situ diffraction

1. Introduction

Mg is ideal for replacing heavier materials in lightweight constructions in the transport sector [1]. Conventional Mg alloys exhibit poor formability at ambient temperatures [2]. Thus, wrought processing is usually carried out above 225 °C, where the activation of non-basal slip becomes possible [3].

The hot deformation mechanisms of metallic alloys are intrinsically related to the stacking fault energy of the material [4]. The distinct role of dynamic recovery (DRV) and dynamic recrystallisation (DRX) [5] promotes a complex microstructure evolution. The reorganisation of dislocation and formation of subgrains or cells occurs during DRV [6]. Furthermore, different types of DRX were proposed, such as discontinuous dynamic recrystallisation (DDRX) [7,8], geometric dynamic recrystallisation (GDRX) [9] and continuous dynamic recrystallisation (CDRX) [5]. They have in common the phenomenon of movement of high angle grain boundaries and the formation of new grains. The nucleation and growth of new grains consuming the deformed material occur during DDRX [10]. GDRX forms refined grains by the impingement of high angle grain boundaries. CDRX develops new HAGBs due to the continuous formation of subgrain boundaries, and their progressive increment in misorientation due to lattice rotation [11].

Mg alloys are typically low stacking fault energy alloys. Thus, limited recovery can lead to the onset of DDRX. The operating deformation mechanisms strongly influenced the role of DRX in ZK60 [12]. Deformation twinning, basal slip and ⟨a + c⟩ dislocation glide was correlated with the role of DRX at temperatures below 200 °C [12]. CDRX occurs due to extensive cross-slip at intermediate temperatures (between 200 °C and 250 °C) [12]. The cross-slip is predominantly activated near original grain boundaries of an *a* type dislocation by the Friedel–Escaig mechanism leading to the transition from a primary screw orientation to an edge orientation [13,14]. A high amount of stacking fault energy of this edge dislocation lies in a non-basal plane [15]. DRX occurred via bulging of grain boundaries and subgrain growth, i.e., DDRX, at temperatures higher than 250 °C [12]. Complementarily, twins with high dislocation density can divide the parent grains leading to the formation of dislocation arrays and low angle grain boundaries within the twins, causing DDRX at higher strains [16]. ZK40 alloy deformed at 350 °C exhibited a similar microstructure evolution to CDRX, evidenced by the subgrain formation, despite forming new grain mostly at grain boundaries [17]. Thus, the identification of the deformation mechanisms and their role on DRX of ZK40 alloy needs to be further clarified.

The Mg-Zn system possesses the potential for the development of low-cost Mg alloys [18]. The enhancement of strength and ductility due to elemental additions, such as rare earth (RE) additions [19,20] occurs through the modification of grain boundary particles [18,21,22]. Among the RE elements, Gd was investigated due to the positive impact on mechanical properties. The creep resistance of Mg-Zn alloys containing Gd improved notably compared to the commercially used WE43 and QE22 alloys [23,24]. Gd can reduce the yield anisotropy (ratio between tensile and compressive yield strength) in extruded Mg alloys [25], and it can also can diminish the strong basal textures [26,27]. The addition of Gd to a ZK40 alloy led to twinning up lo relatively large strains and the formation of a necklace of small grains along grain boundaries via DDRX during hot deformation [17]. Additionally, the addition of Gd to Mg-Zn alloys reduced the segregation of elements in the α-Mg, leading to the increment of DDRX kinetics due to a reduction in solute drag [28]. Thus, the role Gd on the restoration mechanisms of Mg-Zn alloys needs further investigation to tailor the microstructure, and consequently, the mechanical properties of Gd containing Mg alloys.

Ca addition modified the microstructure of Mg alloys [29,30], impacted their mechanical properties [31,32] and promoted texture randomisation [30,33,34]. The addition of CaO forms MgO and Mg_2Ca during casting [35], providing an easier route to add Ca to Mg alloys. Ca addition improved the mechanical properties of extruded and hot-rolled Mg-Zn-Zr alloys due to weaker texture [34]. Ca seemed to retard recrystallisation in Mg-Zn-Zr alloys [36]. The addition of Ca to an AZ80 alloy promoted the formation of a refined recrystallised microstructure and enhanced formability [37].

High energy X-ray diffraction was used for in-situ characterisation of materials undergoing mechanical loading [38–41]. Azimuthal-strain plots bring information on grain size evolution, grain orientation relationships, grain rotation, grain imperfection, and texture evolution [42,43]. The microstructural changes are described in terms of the dislocation slip, twinning, sub-grain formation, recovery and recrystallisation [42].

This work aims to: (a) identify and elucidate the role of the restoration mechanisms during moderate deformation temperatures for ZK40 based alloys; (b) describe the influence of Ca or Gd additions on the microstructure formation and the restoration mechanisms of the investigated alloys. A combined interpretation of in-situ synchrotron radiation diffraction measurements during compression with post-mortem microstructural investigation of the deformed samples provides the insights to describe the deformation mechanisms.

2. Materials and Methods

2.1. Materials

The alloys were prepared with pure Mg, Zn, CaO and master alloys Mg 4 wt.% Gd and Mg-33 wt.% Zr (Zirmax®, Luxfer MEL Technologies, Manchester, UK). Mg was molten in an electric resistance furnace and held at 750 °C, and alloying elements were added to the melt and stirred for 10 min. The melt was then poured into a preheated thin-walled steel mould held at 660 °C for 15 min. Then, the mould was immersed into water at a rate of 10 mm·s^{-1} until the top of the melt was in line with the cooling water. The ingots had a bottom diameter of 250 mm and a height of 300 mm. The indirect casting procedure was adopted to provide a homogeneous microstructure [44]. Table 1 shows the actual compositions of the alloys prepared for this investigation measured with X-ray fluorescence (Zn and Gd) and spark analyser (Ca, Cu, Fe, Ni, Zr). Only Ca is detected in the alloy since CaO dissociates during melting [35]. The alloy with Ca addition is named ZK40-CaO in this work.

Table 1. Chemical compositions of the investigated alloys measured with X-ray fluorescence (Gd and Ca) and spark analyser (Zn, Cu, Fe, Ni, Zr).

Alloys	Zn wt.%	Zr wt.%	Ca wt.%	Gd wt.%	Fe (ppm)	Cu (ppm)	Ni (ppm)
ZK40	5.00	0.53	-	-	11	14	13
ZK40-CaO	4.385	0.34	1.22	-	14	16	14
ZK40-Gd	4.50	0.55	-	1.70	7	29	<30

2.2. In-Situ Synchrotron Radiation Diffraction during Compression

In-situ synchrotron radiation diffraction experiments were performed in reflection mode using the facilities of XTMS (X-ray Scattering and Thermo-Mechanical Simulation) at the Laboratório Nacional de Luz Síncrotron (LNLS, Campinas, Brazil). A monochromatic beam with an energy of 12 keV (λ = 0.10332 nm) and a cross-section of 1.0 × 2.0 mm^2 was used for the current investigation. The diffraction patterns were recorded with a Rayonix® SX 165 detector (Evanston, IL, USA). The sample-to-detector distance of 320 mm and an angle of 32° between the incident and reflected beam were used during the experiments. Only one region of the Debye–Scherrer rings was measured due to the acquisition in reflection mode and the detector size. The range of azimuthal (φ) corresponds to −27.36° a 33.42° while 2θ corresponds to the range between 15.59° to 45.29°.

Specimens with 10 mm in length, 10 mm in width and 5 mm in thickness were used in the in-situ experiments. The illuminated surface of the sample was priorly metallographically prepared up to grinding paper 4000 grit to assure the measured data is as comparable as possible with the bulk material. The specimens were placed in the chamber of a Gleeble® 3S50 (Gleeble, Poestenkill, NY, USA). The specimens were heated up to the deformation temperature at 10 °C s^{-1} and held at the testing temperature for 3 min before deformation to ensure temperature homogeneity. The deformation was carried at 250 °C and 300 °C, controlled using a K-type thermocouple welded at the surface of the specimen. The specimens were compressed up to 0.3 of the true strain at 0.0001 s^{-1} and 0.001 s^{-1}. A protective Ar atmosphere was used, and the deformation was followed by a final Ar quenching. The details of the experimental setup can be found in [45].

The Debye–Scherrer rings were analysed using software Dracon (Diffracted X-rays Analysis Console, LNLS, Brazil) and the ImageJ® software package [46]. The two-dimensional diffraction patterns were converted into azimuthal angle–time/strain plots to study the evolution of the microstructure, according to the methodology explained in [47] which is briefly summarised: (i) stacking of the recorded 2D images, (ii) selection of the diffraction ring and conversion into cartesian coordinates with a final 3D volume of axis 2θ, t (or strain ε), and φ (azimuthal angle) and (iii) projection over the t (or ε)-φ plane. The azimuthal angle–time/strain plots were generated for the chosen crystallographic planes.

2.3. Microstructure Characterisation

The compressed samples were cold mounted, ground using SiC paper and polished using OPS neutral solution. The specimens analysed with optical microscopy (OM) were etched with an acetic picric acid solution [2]. The microstructure was analysed using a reflected light microscope Leica DMI 5000 (Leica, Wetzlar, Germany). The AnalySIS Pro software (Camo Analytics, Oslo, Norway) and its extensions were used to determine the grain size using the intercepts method according to the standard ASTM E112-12.

Samples without etching were analysed with scanning electron microscopy (SEM) using a Zeiss FEG-SEM Ultra 55 (Zeiss, Oberkochen, Germany) and a TESCAN Mira3 (Tescan, Brno, Czechia) electron scanning microscopes, both equipped with energy-dispersive X-ray spectroscopy (EDXS) microanalysis hardware as well as with a Hikari camera and a TSL-OIM Data Collector (EDAX, Mahwah, NJ, USA) software package to perform electron backscattered diffraction (EBSD) measurements. A voltage of 15 kV, a working distance of 15 mm, and a spot size of 5 nm were used for acquiring backscattered electron (BSE) images and for EDXS analysis. A minimum of five representative BSE micrographs was analysed using the software ImageJ® [46] to determine the area fraction of intermetallic compounds in each alloy. A voltage of 30 kV, a working distance of 25 mm and spot size of 80 nm were used for the EBSD measurements of the deformed samples carried out in areas of 500 μm × 500 μm close to the centre of each specimen, with a step size of 0.5 μm. The software OIM Analysis v.8 (EDAX, Mahwah, NJ, USA) was used for data treatment. A misorientation angle of 15° was used to define a high angle grain boundary, and a minimum grain size of 2 μm was selected. The grains were standardised concerning their confidence index (CI), and a minimum CI of 0.2 was used to clean the data with respect to the neighbour grains. A maximum of 10% of the measured points was cleaned. All microstructures are presented with the compression direction vertical to the page. Tensile twins $\{10\bar{1}2\}\langle\bar{1}011\rangle$ were calculated using a direction tolerance angle of 10° and a K1 plane tolerance angle of 3°. Texture analysis was performed using the harmonic series expansion method with a series rank of 16, a Gaussian smoothing of 5°, and a triclinic sample symmetry.

3. Results

The as-cast and deformed microstructures are analysed as well as the in-situ synchrotron radiation diffraction measurements.

3.1. As-Cast Microstructures

Light optical and BSE micrographs, as well as EDXS line scans, are used to characterise the as-cast microstructure of the investigated alloys, Figure 1. The light optical micrographs (Figure 1a–c) shows equiaxed-like grains for the investigated alloys. The average grain size of the ZK40 was 72.4 ± 2.5 μm, 75.9 ± 5.0 μm for the ZK40-CaO. The coarse and fine-grained ZK40-Gd gives rise to two different grain distribution with a global average of 46.1 ± 9.1 μm. The average size of the fine grains is 30.1 ± 4.2 μm, while coarse grains were 81.9 ± 9.2 μm. Intermetallic compounds are formed at grain boundaries for the ZK40 alloy, Figure 1d. A near-continuous network of intermetallic compounds is formed along the grain boundaries for the ZK40-CaO alloy, Figure 1e. Similarly, the ZK40-Gd exhibits a near-continuous network of intermetallic compounds along the grain boundaries, Figure 1f. The volume fraction of second phases is 1.6 ± 0.5%, 6.5 ± 0.9%, and 5.7 ± 1.0% for ZK40, ZK40-CaO and ZK40-Gd, respectively. The ZK40 alloy contains $MgZn_2$ phase, as shown in [48,49]. The ZK40-CaO and ZK40-Gd contain $Ca_2Mg_6Zn_3$

and (Mg, Zn)$_3$Gd$_2$, respectively [48,49]. Figure 1g–i shows the EDXS line scans for the investigated alloys. In ZK40 and ZK40-Gd, Zn segregates and its concentration is higher near grain boundaries. Zn segregation is near neglectable in ZK40-CaO. Inverse segregation of Zr occurs in ZK40-CaO and ZK40-Gd. A small amount of Ca and Gd are detected in the α-Mg matrix of the ZK40-CaO and ZK40-Gd, respectively. However, no appreciable segregation of Ca or Gd is observed. A more detailed microstructure investigation of the as-cast alloys used in this study can be found in [48,49].

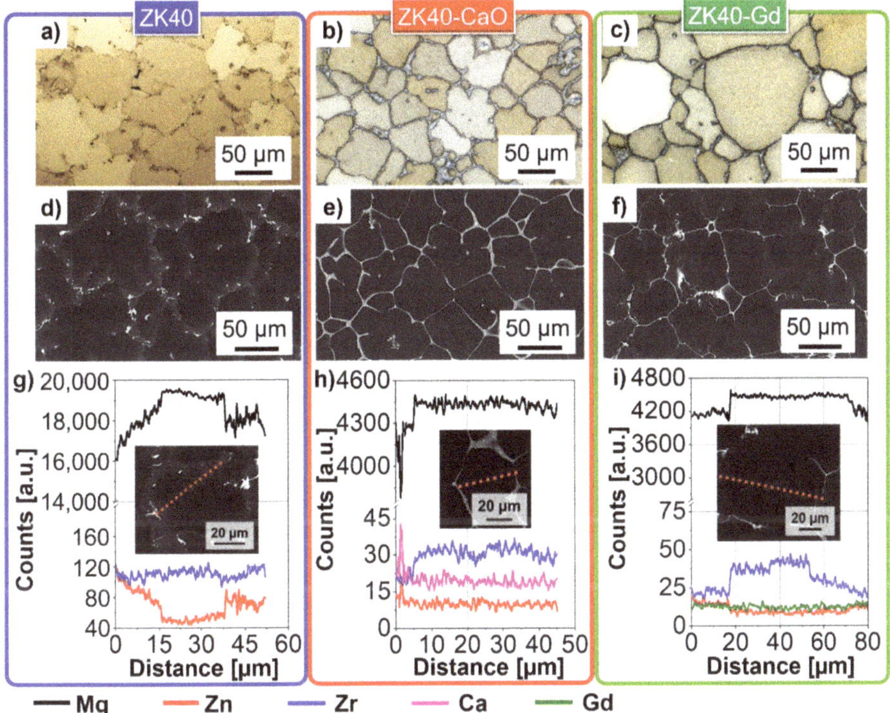

Figure 1. (**a–c**) Light optical micrographs, (**d–f**) backscattered electron (BSE) micrographs, and (**g–i**) energy dispersive X-ray spectroscopy (EDXS) line scans of the as-cast alloys: (**a,d,g**) ZK40; (**b,e,h**) ZK40-CaO; (**c,f,i**) ZK40 Gd.

3.2. Flow Curves

The measured flow curves at 250 °C and 300 °C are shown in Figure 2a,b, respectively. ZK40-Gd and ZK40-CaO exhibit similar behaviour at 250 °C and 0.001 s^{-1}. A fast work hardening occurs once the plastic regime is reached, followed by a progressive decrease in work hardening until a plateau at a true strain of ~0.2. This plateau does not necessarily correspond to the steady-state condition that can be achieved at larger strains. Hradilová et al. [50] investigated alloys of the similar chemical composition of the ZK40-CaO and showed that DDRX could take place at strains larger than 0.3, leading to flow softening. The ZK40 exhibits less pronounced work hardening and lower flow stresses than ZK40-CaO and ZK40-Gd, although the plateau is also reached at a true strain of ~0.2. A decrease in the strain rate to 0.0001 s^{-1} leads to lower flow stresses and very low work hardening for all alloys. Once the plastic deformation starts, a rapid work hardening is followed by a plateau at strains lower than 0.1. The ZK40 and ZK40-Gd exhibit similar behaviour at 250 °C and 0.0001 s^{-1}, while ZK40-CaO exhibits slightly higher flow stress.

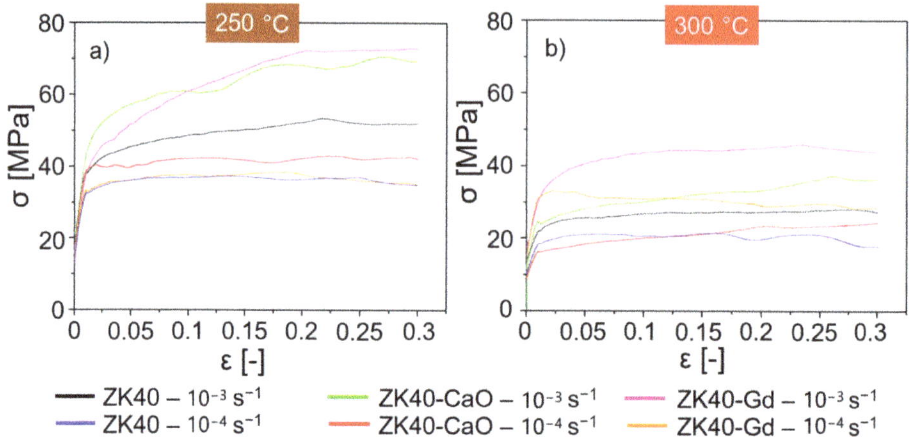

Figure 2. Flow stress for the alloys tested at (**a**) 250 °C and (**b**) 300 °C.

The increase in temperature to 300 °C leads to a pronounced reduction in the maximum stress for the investigated alloys (~50% of the respective ones at 250 °C), Figure 2b. The following occurs for deformation at 300 °C and 0.001 s^{-1}: ZK40 and ZK40-Gd show fast work hardening and reach a plateau at low strains (~0.1), while ZK40-CaO shows a nearly constant but minimal work hardening. The decrease of strain rate to 0.0001 s^{-1} leads to three different behaviours: (a) ZK40 shows a rapid work hardening followed by a plateau at a true strain of ~0.05; (b) ZK40-CaO shows similar flow stresses compared to ZK40 but a nearly constant work hardening; (c) After a rapid work hardening, an almost continuous flow softening occurs for the ZK40-Gd.

3.3. Deformed Microstructures

The microstructure of the Mg-matrix of the deformed ZK40 up to 0.3 strain is shown in Figure 3. The inverse pole figure (IPF) maps (Figure 3a–c) show that the formed microstructural features are comparable: pronounced basal texture, a network of low angle grain boundaries indicated by white lines and misorientation spread within the grain. The pronounced basal texture is also visible in the pole figures in Figure 3c–f. A slightly more isotropic deformation seems to occur at 300 °C and 0.0001 s^{-1}, as indicated by the smallest maximum texture index and more random distribution of poles in the pole figure.

Figure 4 shows the EBSD results for the deformed ZK40-CaO up to 0.3 strain. Similar to the ZK40, the formation of low-angle grain boundaries was more concentrated in regions near the grain boundaries, while intensive misorientation spread is present in nearly all grains. The maximum index of texture is slightly higher than that for the ZK40. The deformation seems to be less anisotropic at 300 °C and 0.0001 s^{-1}.

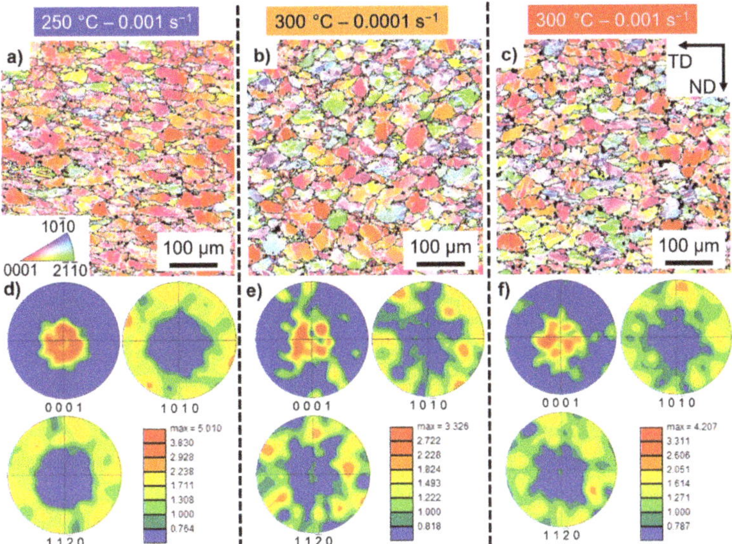

Figure 3. Electron backscattered diffraction results for the ZK40 alloy: (**a**–**c**) inverse pole figure maps and (**e**,**f**) pole figures. The tested conditions: (**a**,**d**) 250 °C and 0.001 s^{-1}; (**b**,**e**) 300 °C and 0.0001 s^{-1}; (**c**,**f**) 300 °C and 0.001 s^{-1}. Only α-Mg is indexed. Non-indexed regions are highly deformed regions or intermetallic compounds. White and black lines indicate low and high angle grain boundaries, respectively. ND and TD indicate the compression and transversal directions, respectively.

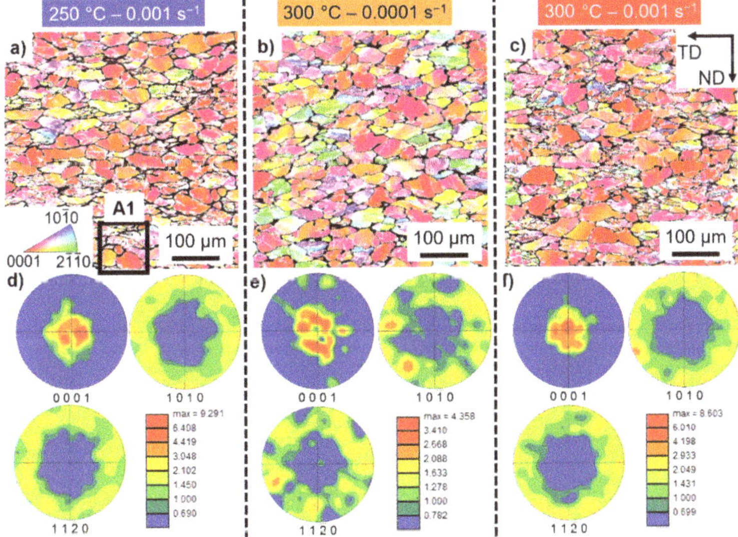

Figure 4. Electron backscattered diffraction results for the ZK40-CaO alloy: (**a**–**c**) inverse pole figure maps and (**e**,**f**) pole figures. The tested conditions: (**a**,**d**) 250 °C and 0.001 s^{-1}; (**b**,**e**) 300 °C and 0.0001 s^{-1}; (**c**,**f**) 300 °C and 0.001 s^{-1}. Only α-Mg is indexed. Non-indexed regions are highly deformed regions or intermetallic compounds. White and black lines indicate low and high angle grain boundaries, respectively. Area A1 is shown in Figure 14. ND and TD indicate the compression and transversal directions, respectively.

Figure 5 shows the deformed microstructure of the α-Mg matrix for the ZK40-Gd up to 0.3 strain. The microstructure is comparable to the ZK40 and ZK40-CaO consisting of an intricated network of low angle grain boundaries, a high level of misorientation spread within the α-Mg and intense basal texture. The isotropic degree of deformation is higher at 300 °C than at 250 °C, as grains of different orientations are depicted. Besides, the pole figures show a more diffuse basal texture at 300 °C and poles in other directions than the basal one, Figure 5e,f. The most isotropic condition was the 300 °C and 0.001 s^{-1} for the ZK40-Gd, instead of 300 °C and 0.0001 s^{-1} observed for ZK40 and ZK40-CaO.

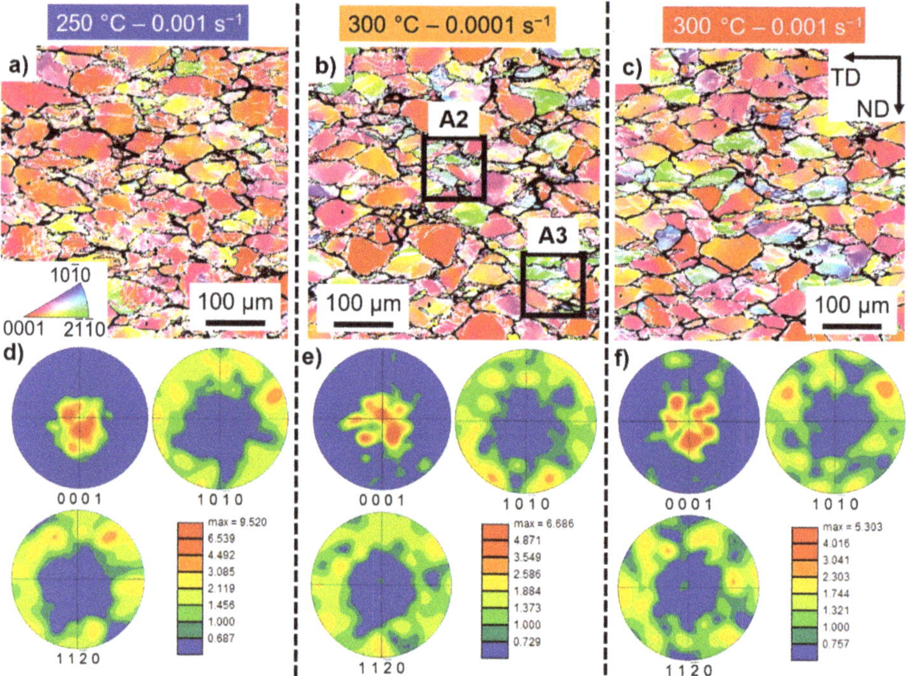

Figure 5. Electron backscattered diffraction results for the ZK40-Gd alloy: (**a**–**c**) inverse pole figure maps and (**e**,**f**) pole figures. The tested conditions: (**a**,**d**) 250 °C and 0.001 s^{-1}; (**b**,**e**) 300 °C and 0.0001 s^{-1}; (**c**,**f**) 300 °C and 0.001 s^{-1}. Only α-Mg is indexed. Non-indexed regions are highly deformed regions or intermetallic compounds. White and black lines indicate low and high angle grain boundaries, respectively. Areas A2 and A3 are shown in Figure 14. ND and TD indicate the compression and transversal directions, respectively.

Figure 6 shows the quantification of the overall microstructural features for the investigated conditions. The boundary density (S_v) is calculated as the total line length of each boundary divided by the measurement area and is shown in Figure 6a–c for the ZK40, ZK40-CaO and ZK40-Gd, respectively. Higher S_v values of low angle grain boundaries are observed at 250 °C compared to 300 °C. Slightly higher S_v values of low angle grain boundaries are measured at 0.001 s^{-1} than 0.0001 s^{-1} for the ZK40 and ZK40-CaO. On the other hand, ZK40-Gd shows slightly lower values of low angle grain boundaries at 250 °C and 0.001 s^{-1} than those at 250 °C and 0.0001 s^{-1}. ZK40 shows the highest values of S_v of low angle high angle grain boundaries, while comparable values are found for ZK40-CaO and ZK40-Gd.

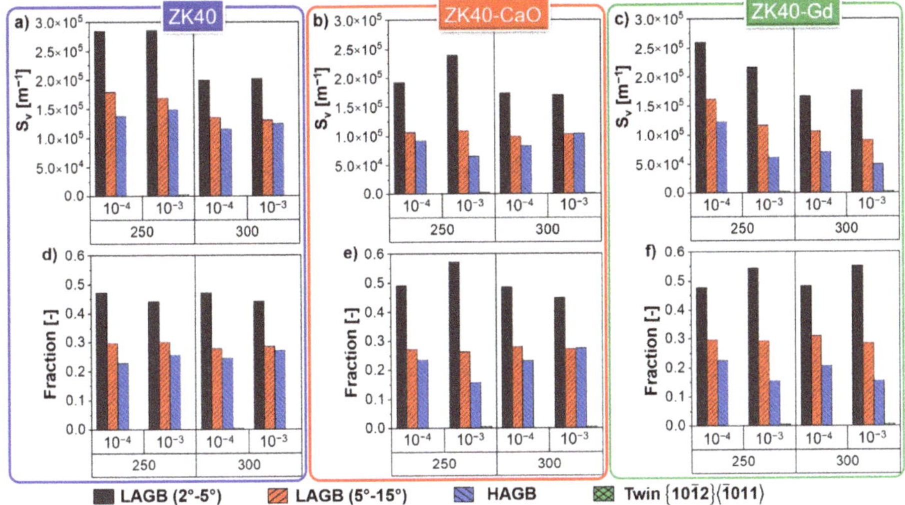

Figure 6. Microstructural features evaluated from the electron backscattered diffraction (EBSD) measurements for the (**a**,**d**) ZK40; (**b**,**e**) ZK40-CaO; and (**c**,**f**) ZK40-Gd. Boundary density (S_v) (**a**–**c**) and fraction of boundaries (**d**–**f**) are shown.

The initial grain size indicates the initial value of the boundary density of high angle grain boundaries. Smaller grains correspond to higher boundary density. The initial grain size of ZK40 and ZK40-CaO are comparable, while ZK40-Gd shows a slightly smaller overall grain size. ZK40-Gd deformed at 300 °C, and 0.001 s^{-1} shows the smallest value of S_v for the high angle grain boundaries, 5×10^4 m^{-1}. ZK40 deformed at 250 °C, and 0.001 s^{-1} shows the highest value, 15×10^4 m^{-1}. The three times the higher value for the ZK40 indicates that formation of high angle grain boundaries occurs during deformation in this alloy.

Figure 6d–f shows the fraction of each boundary type for the ZK40, ZK40-CaO and ZK40-Gd, respectively. The low angle grain boundaries in the range between 2° and 5° are the predominant boundary type for all investigated conditions. ZK40 shows the highest fraction of high angle grain boundaries, except at 300 °C and 0.001 s^{-1}, where the fraction of high angle grain boundaries is comparable with that for ZK40-CaO. ZK40 shows similar values of fraction of low angle grain boundaries in the range between 2° to 5° for the investigated conditions. ZK40-Gd deformed at 0.001 s^{-1} shows a higher fraction of low angle grain boundaries in the range between 2° and 5° compared to 0.0001 s^{-1}. ZK40-CaO shows a non-linear dependency on temperature and strain rate.

The plastic deformation occurs in the α-Mg matrix, while the brittle intermetallic compounds fragments during deformation, as shown in Figure 7. The intermetallic compounds present in the ZK40 are isolated and in low volume fraction. They are not shown here because they do not exhibit the brittle behaviour like the intermetallic compounds present in the modified alloys. The intermetallic compound formed in the ZK40-Gd has a typical lamellar structure [48]. Its fracture leads to the formation of fragmented fine particles of that phase, as indicated by the dashed red circle in Figure 7d. The intermetallic compound formed in the ZK40-CaO has a mixed structure of platelet and fine lamellar structure [49]. The red arrows in Figure 7a,b indicate the cracks that fragment the network of intermetallic compounds into smaller particles.

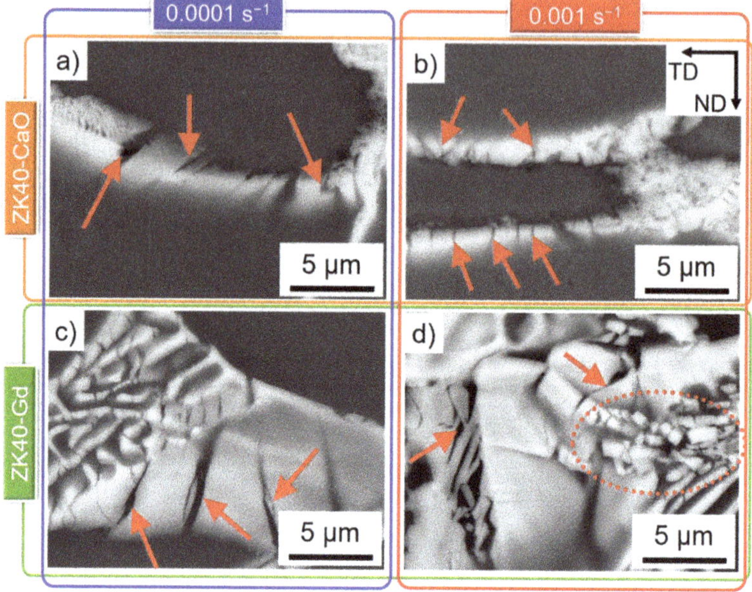

Figure 7. Backscattered electron micrographs of the fractured intermetallic compounds in: (**a**,**b**) ZK40-CaO; and (**c**,**d**) ZK40-Gd, deformed at 0.001 s^{-1} and: (**a**,**c**) 250 °C; and (**b**,**d**) 300 °C. Red arrows indicate cracks. The fragmented intermetallic compound is indicated by the dashed red ellipse. ND and TD indicate the compression and transversal directions, respectively.

3.4. Insights from In-Situ Synchrotron Radiation Diffraction

The azimuthal-strain plots of the $\{10\bar{1}0\}$ and {0002} are shown in Figure 8 for the deformed samples at 0.0001 s^{-1}. The acquisition of the Debye-Scherrer rings in reflection mode, and the detector size limits the range of azimuthal from −22° to 27°. The deformation starts with an intense spread of orientation in the diffraction rings and the decay in the intensity of the diffraction signal, interpreted as the formation of misorientation within α-Mg because of rapid formation of dislocations due to glide, i.e., work hardening. This process is indicated in Figure 8 by the blue arrows. Deformation twinning can also occur at the early stages of deformation. However, it is not straightforward to be depicted by the obtained azimuthal-strain-plots. The result is the progressive disappearance of the Bragg spots in the azimuthal range shown in Figure 8.

The bending of the timelines is a second phenomenon interpreted from azimuthal-time plots, as indicated by the black arrows in Figure 8. The bending of the timelines shows that the plastic deformation leads to lattice rotation in some grains that bends the grain orientation in the azimuthal-time plots. The progressive disappearance of those timelines indicates those grains are not in Bragg position as a consequence of the pronounced basal texture formation, as shown in Figures 3–5. Dynamic recovery causes the thinning of the timelines due to the reorganisation of dislocations, diminishing the local misorientation spread. Finally, the general spread in orientation and decay in intensity of the timelines occurs due to the ongoing formation of misorientation within the grains as a mechanism of plastic deformation accommodation.

Despite the comparable azimuthal-strain plots at 250 °C in Figure 8 for the investigated conditions, there is a significant difference between 250 °C and 300 °C: new spots are observed at 300 °C, as highlighted in the magnified area, and is more pronounced for the ZK40. The spotty characteristic of the azimuthal-strain plots can occur due to:

- Twinning: the progressive appearance of a timeline in a crystallographic plane with the simultaneous and progressive disappearance of the timeline corresponding to the parent grain in another crystallographic plane.
- DDRX: nucleation and growth of new grains that appear in the gauge volume and as new timelines.

The observed new spots and timelines likely correspond to the formation of new grains since they appear as a thinning of diffuse regions, and those crystals rotate notably. Their subsequent disappearance can be attributed to:

- Sufficiently high local plastic deformation leads to lattice rotation of the formed grains, consequently, to the disappearance of their corresponding timelines.
- A fraction of the new grains can be pushed away from the gauge volume. Oppositely, new ones can enter the gauge volume, leading to the appearance of new spots.

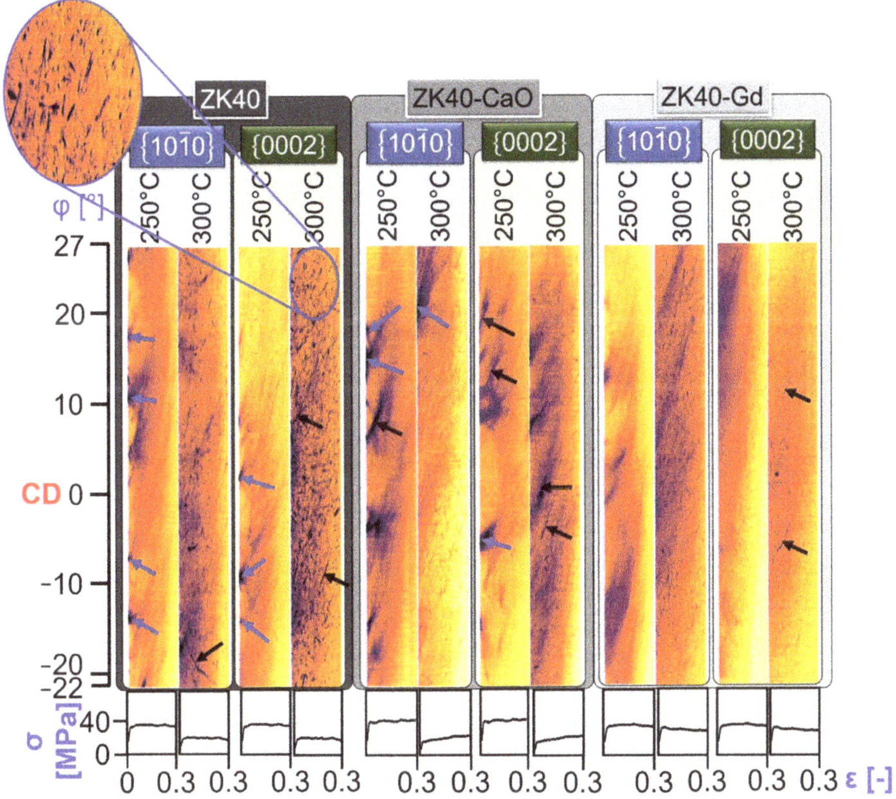

Figure 8. Azimuthal-strain plots for the three investigated alloys tested at 250 °C and 300 °C for the strain rate of 0.0001 s^{-1}. Blue arrows indicate the intense spread of orientation in the diffraction rings and the decay in the intensity of the diffraction signal. Black arrows indicate the bending of the timelines. CD indicates the compression direction.

The azimuthal-time plots of the of the $\{10\bar{1}0\}$ and {0002} are shown in Figure 9 for the deformed samples at 0.001 s^{-1}. Similar to the azimuthal-strain plots obtained for the 0.0001 s^{-1} (Figure 8), the deformation of the investigated alloys at 0.001 s^{-1} leads to the formation of:

- Misorientation spread: the intense spread of orientation in the diffraction rings and the decay in the intensity of the diffraction signal.
- Crystal and subgrain rotation: bending of the timelines.

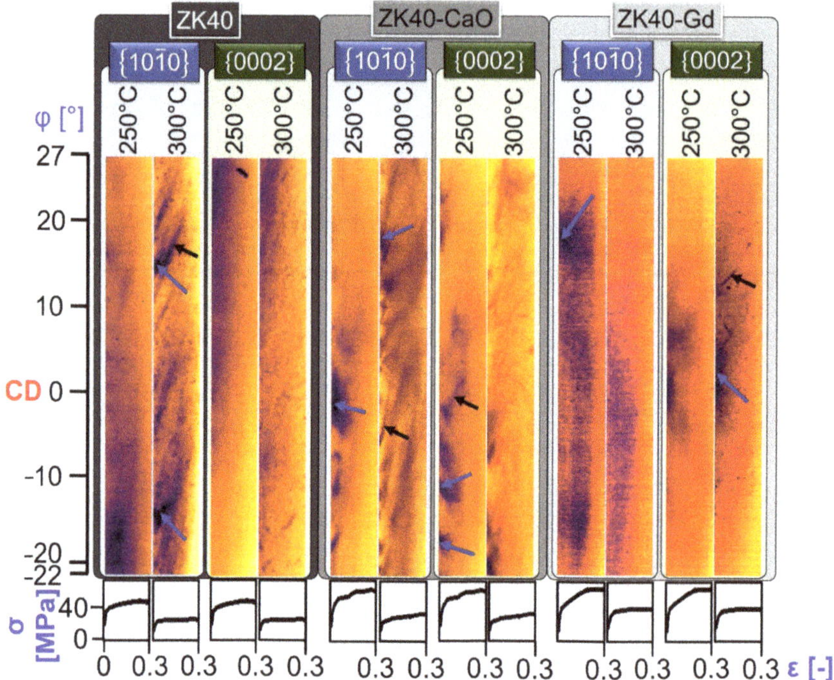

Figure 9. Azimuthal-strain plots for the three investigated alloys tested at 250 °C and 300 °C for the strain rate of 0.001 s^{-1}. Blue arrows indicate the intense spread of orientation in the diffraction rings and the decay in the intensity of the diffraction signal. Black arrows indicate the bending of the timelines. CD indicates the compression direction.

The spread of crystal orientation and the disappearance of the timelines show the first phenomenon. The blue arrows in Figure 9 and the diffuse formation timelines in Figure 9 indicate the misorientation spread. The black arrows in Figure 9 indicate the second phenomenon. The timelines in the azimuthal-strain plots for 0.001 s^{-1} are more diffuse than those of the deformation at 0.0001 s^{-1} (Figure 8) because of:

- The stack of less Debye–Scherrer rings: 10% of the ones used for the 0.0001 s^{-1}
- Higher dislocation density and higher misorientation spread within the grains formed at higher strain rates.

4. Discussion

The in-situ and ex-situ results for the deformation at moderate temperatures of ZK40 alloy and modified ones with individual addition of Ca and Gd show that dynamic recovery is the primary restoration mechanism of the α-Mg up to the investigated strain. Simultaneously, the brittle behaviour of the intermetallic compounds leads to their fragmentation and, consequently, flow softening. Figures 3–5 elucidate the role of dynamic recovery, where the formation of a substructure with low angle grain boundaries is present for all conditions. Figure 6 shows the significant fraction of low angle grain boundaries in the range between 2° to 5° that are typically cell or subgrain boundaries formed by the reorganisation of dislocations via dynamic recovery. Their fraction is the highest among the measured boundary types in all investigated cases. The formed boundaries between 5° and 15° are correlated to a higher level of local plastic deformation, leading to a higher degree of local lattice rotation. The plastic deformation is accommodated either by the formation of misorientation spread through the α-Mg matrix, slip bands, and formation of a band-like structure or by dynamic recovery and organised boundaries. The plateau that follows the work hardening in the flow stresses in Figure 2 occurs when the consumption of dislocation via dynamic recovery balances their production.

However, dynamic recovery does not occur homogeneously within the α-Mg grain. Figure 10 shows the kernel average misorientation analysis for the deformed alloys at 300 s^{-1} and 0.001 s^{-1}. The kernel average misorientation distribution, Figure 10d shows nearly the same distributions for the three alloys. The maps, though, shows that the formed boundaries distribute more homogeneously for the ZK40 alloy. At the same time, the new boundaries localise in the vicinity of intermetallic compounds and within individual grains for the ZK40-CaO. Thus, Ca addition seems to hinder the reorganisation of dislocations into low angle grain boundaries. ZK40-Gd shows an intermediate behaviour, where a more homogeneous substructure is formed compared to the ZK40-CaO, but less homogeneous compared to the ZK40.

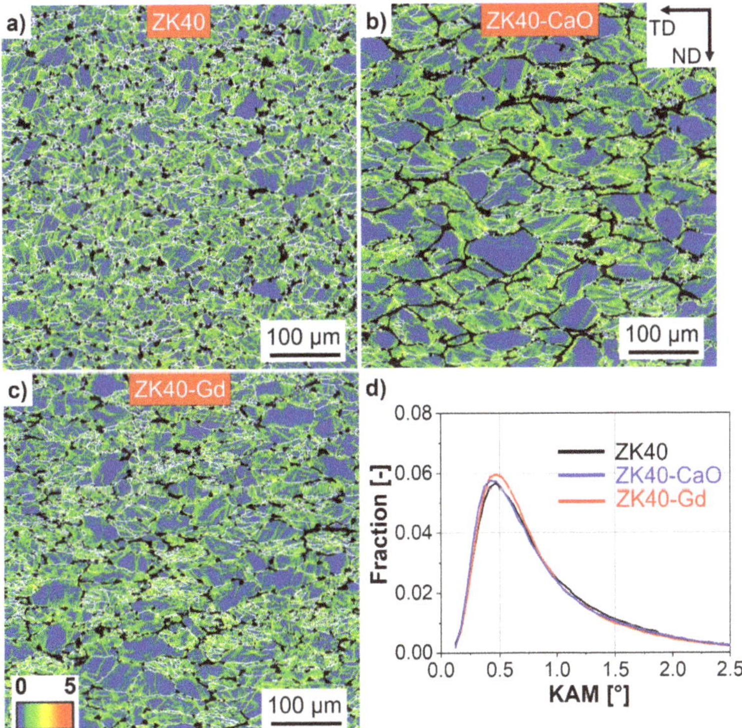

Figure 10. Kernel average misorientation maps for the alloys deformed at 300 °C and 0.001 s^{-1}: (**a**) ZK40, (**b**) ZK40-CaO and (**c**) ZK40-Gd. Kernel average misorientation distributions for the three alloys showed in (**d**). Only α-Mg is indexed. Non-indexed regions are highly deformed regions or intermetallic compounds. White lines indicate high angle grain boundaries. ND and TD indicate the compression and transversal directions, respectively.

Dynamic recovery is the major but not the only restoration mechanism for the investigated deformation conditions. Complementary to Figures 3–5 and 11 shows inverse pole figure maps in higher resolution of the deformed alloys at 0.001 s^{-1}. Despite the comparable overall microstructural features (Figures 3–5) for investigated alloys and small differences concerning the different deformation conditions, the differences are notable on the local scale. Black ellipses highlight the presence of microstructural features that resembles deformation twins. The presence of misorientation spread, different orientations and boundaries within those regions indicate that if they were deformation twins, they were formed in an early stage of deformation followed by further local lattice rotation and dynamic recovery. Hradilová et al. [50] showed similar microstructural features for Mg-Zn-Ca alloys deformed in a similar temperature range. They explained those features as the formation of tensile twins followed by accumulation of dislocations at the twin boundaries, dynamic recovery and the formation of boundaries within the twins. Finally, the higher stored energy at twins [51] leads to the bulging of new grains [50]. The first slope of work hardening observed in Figure 2a and more evident for the ZK40-CaO and ZK40-Gd can correspond to the formation of twins, while the second slope of work hardening before achieved plateau-like flow stress can correspond to twin growth and increment of dislocation density within them. The negligible presence of twins at the post-mortem microstructure at 0.3 strain agrees with the observation of Barnett et al. [52] that twinning is unlikely to occur if slip starts to dominate at larger strains. Twinning as a predominant mechanism at the beginning of deformation is also predicted due to the relatively large initial grain size [53,54] for the investigated alloys. Therefore, DRX at 0.001 s^{-1} seems to occur due to the nucleation and growth of new grains via DDRX at twinned regions at the early stages of deformation and is more pronounced for the ZK40 alloy.

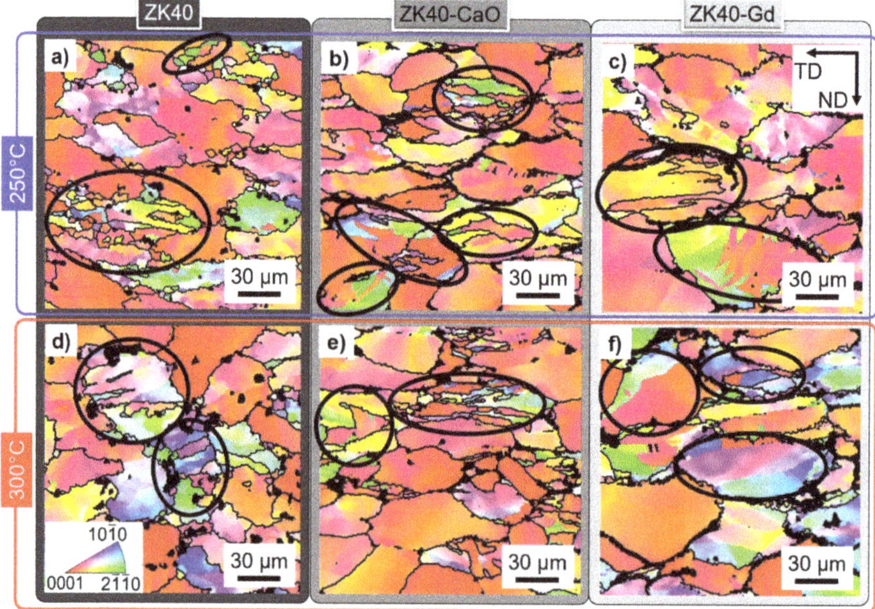

Figure 11. Inverse pole figure maps for the deformed alloys at 0.001 s^{-1}: (**a**,**d**) ZK40; (**b**,**e**) ZK40-CaO; (**c**,**f**) ZK40-Gd; and: (**a**–**c**) 250 °C, (**d**–**f**) 300 °C. Only α-Mg is indexed. Non-indexed regions are highly deformed regions or intermetallic compounds. Black lines indicate high angle grain boundaries. Black ellipses indicate regions that resemble twins. ND and TD indicate the compression and transversal directions, respectively.

The azimuthal-strain plots in Figure 8 show the presence of new spots. It is clarified by the formation of new grains at 0.0001 s^{-1}, as shown in Figure 12. Two mechanisms are proposed:

- Twins that recover and promote nucleation of new DDRX grains as explained by Hradilová et al. [50] and indicated by the blue ellipses in Figure 12c.
- Pile-up of dislocation along grain boundaries and along intermetallic compounds leading to bulging of new grains via DDRX, as indicated by the black arrows in Figure 12, followed by growth as indicated by the white arrow arrows in Figure 12.

The azimuthal-strain plots in Figure 8 show that new spots are evident at 300 °C and negligible at 250 °C. The comparison between Figure 12a,b clarifies the different DRX mechanisms for the ZK40 at 250 °C and 300 °C, respectively. Despite a small fraction of DDRX grains observed at 250 °C, as indicated by the black arrows in Figure 12a, their presence is notably more pronounced at 300 °C. At 250 °C, the formation of new grains occurs via:

- The formation of a substructure followed by CDRX, as evidenced in the azimuthal-strain plots, Figure 8: firstly, the intense spread of orientation in the diffraction rings and the decay in the intensity of the diffraction signal is associated with the misorientation spread and the multiplication of dislocations. It is followed by the bending and thinning of the timelines associated with subgrain rotation and dynamic recovery, respectively. The white ellipses in Figure 12a indicate regions where subgrains of different orientations are found within a grain. As proposed by Galiev et al. [12], extensive cross-slip at lower temperature leads to dislocation rearrangements into low angle grain boundaries networks, followed by CDRX.
- Bulging of new grains via DDRX indicated by the black arrows in Figure 12a.

Thus, higher temperatures promote DDRX for the ZK40 due to higher boundary mobility and higher dislocation climb activity, promoting bulging of the deformed grain boundaries and the growth of new grains. The controlling mechanisms are, though, a function not only of temperature but also of strain. Moreover, different mechanisms can co-occur, as also observed by Galiyev et al. [12].

The addition of Gd or Ca also has an impact on the DRX kinetics. The presence of recrystallised grains is more pronounced for the ZK40 (Figure 12a,b), while Gd addition leads to the formation of a lower amount of new grains at 300 °C and 0.0001 s^{-1} up to the strain of 0.3, as indicated in Figure 12d. On the other hand, the presence of new grains at 300 °C and 0.0001 s^{-1} seems negligible for the ZK40-CaO. Hradilová et al. [50] showed that the peak stress is higher or equal to 0.3 for alloys of similar composition. Following the assumption proposed by Galiyev et al. [12] for a ZK60 alloy, Ca seems to modify the diagram of controlling mechanisms so that each field shifts to higher strains and higher temperatures. Thus, Ca additions promote twinning, reduces climb and bulging of new grains, retarding DDRX in Mg-Zn alloys.

Gd also seems to hinder DDRX. Despite the slightly smaller initial grain size than the ZK40 (Figure 1), the boundary density of high angle grain boundaries is smaller for the deformed ZK40-Gd up to 0.3 of a strain than the ZK40 (Figure 6). This result seems contradictory with the finding of Hoseini-Athar et al. [28], which showed that the addition of Gd to Mg-Zn alloys reduces the segregation of elements in the α-Mg, leading to the increment of DDRX kinetics due to a reduction in solute drag.

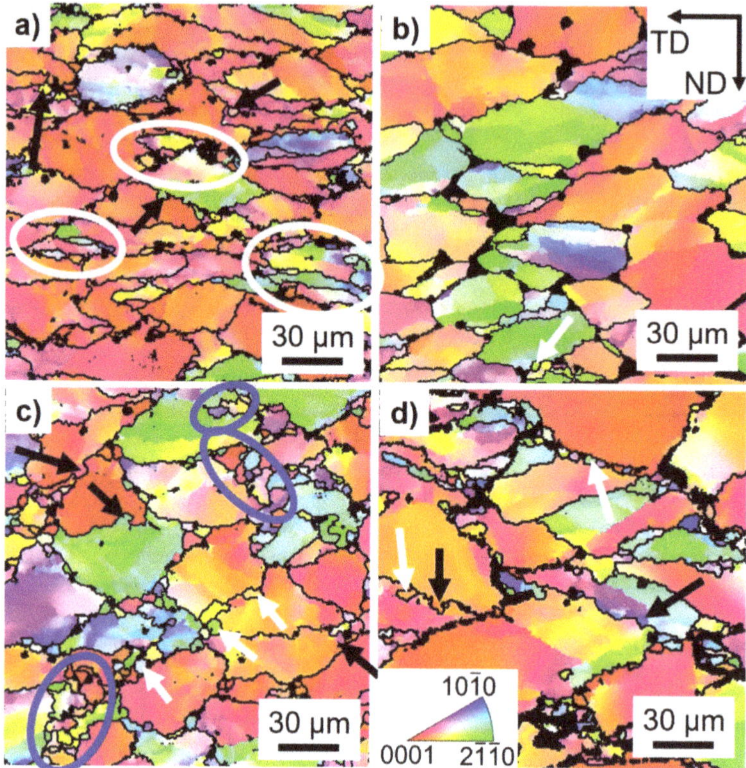

Figure 12. Inverse pole figure maps for the deformed alloys at 0.0001 s^{-1} and: (**a**) 250 °C, (**b–d**) 300 °C; for the: (**a**,**b**) ZK40; (**c**) ZK40-CaO; (**d**) ZK40-Gd. Only α-Mg is indexed. Non-indexed regions are highly deformed regions or intermetallic compounds. Black lines indicate high angle grain boundaries. Black arrows indicate small grains at prior grain boundaries. Blue ellipses indicate regions that resemble a twin shape and also new grains. ND and TD indicate the compression and transversal directions, respectively.

Figure 1g–i shows that small content of Ca, and Gd is present in the α-Mg matrix and that the segregation of Zn and Zr is pronounced. The ZK40 alloy shows mostly higher segregation of Zn. Figure 13 shows the EDXS maps measured simultaneously with EBSD for the Zn segregation in the deformed samples at 300 °C and 0.0001 s^{-1}. The segregation of Zn is still present after deformation and more pronounced for the ZK40, Figure 13a. While the investigated alloys by Hoseini-Athar et al. [28] contained isolated intermetallic particles, the investigated ZK40-Gd has a semi-continuous network of intermetallic compounds along the grain boundaries. Due to the higher strength, the network of $(Mg,Zn)_3Gd_2$ withstand an elevated part of the stress until the network starts to fragment (Figure 7). Then, the load is progressively transferred into the α-Mg, leading to flow softening [17], as shown in Figure 2. Therefore, until the semi-continuous network of $(Mg,Zn)_3Gd_2$ withstands the load, the plastic deformation is limited in the α-Mg matrix. Furthermore, the α-Mg deforms near the intermetallic compounds, evident by the formation of boundaries preferentially along grains boundaries. Therefore, the formation of a semi-continuous network of $(Mg,Zn)_3Gd_2$ reduces the role of DDRX in ZK40-Gd.

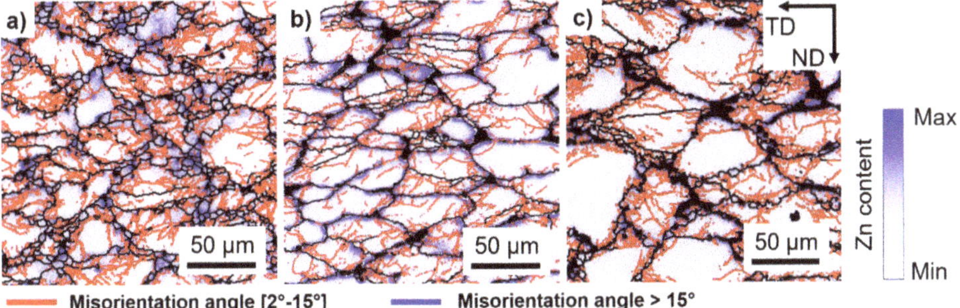

Figure 13. Energy-dispersive X-ray spectroscopy (EDXS) maps of Zn normalised for de deformed alloys at 300 °C and 0.0001 s^{-1}: (**a**) ZK40, (**b**) ZK40-CaO, (**c**) ZK40-Gd. Only α-Mg is indexed. Non-indexed regions are highly deformed regions or intermetallic compounds. Low and high angle grain boundaries are indicated by red and black lines, respectively. ND and TD indicate the compression and transversal directions, respectively.

The flow stress behaviour in Figure 2b and the formation of new grains are directly correlated: the ZK40 and the ZK40-Gd exhibit a flow softening due to the formation of recrystallised grains, while ZK40-CaO exhibits a slight work hardening, where the formation of new grains is nearly absent. The fragmentation of the intermetallic compounds, as shown in Figure 7, seems to contribute to an additional softening for the ZK40-Gd. However, the fragmentation of the intermetallic compounds seems to play a minor role in the flow stress evolution for the ZK40-CaO.

Finally, the mechanisms of microstructure formation and the role of dynamic recovery are derived from the interpretation of the microstructure features, Figure 14. Figure 14a,d,g shows a notable misorientation spread in the inverse pole figure maps shown as a progressive change in orientation within each grain. Besides, Figure 14c illustrates the misorientation profile I with variations in misorientation without the formation of boundaries. The kernel average misorientation map (Figure 14b) also indicates diffuse boundaries within the observed grain, indicating an early degree of the reorganisation of dislocations via dynamic recovery. The misorientation profile II in Figure 14f shows a region that resembles the morphology of a twin. The boundary misorientation is far from 90° (typically for the tensile deformation twins), indicating that the feature cannot be classified as a twin. However, it suggests that it corresponds to a twin formed at early stages of deformation that further deformed forming misorientation spread and boundaries within the twinned region, as proposed by Hradilová et al. [50]. The kernel average misorientation map in Figure 14e shows a diffuse pattern within a parent un-twined grain, indicating the parent grain also accommodates further plastic deformation.

The increase in misorientation spread within α-Mg occurs due to lattice rotation caused by the formation of geometrically necessary dislocations or the formation of LAGBs [42]. On the other hand, the rotation of the subgrains with lower Taylor factor (softer subgrains) withstands larger plastic deformation. Both phenomena indicate DRV as the predominant mechanism for the microstructure evolution of the α-Mg, corroborating with the interpretation of the above-observed microstructures. Complementary, Figure 14g shows a grain with a band-like structure that resembles twins. The corresponding misorientation profile in Figure 14i shows that the boundary misorientations are smaller than 40°. The band-structure, thus, is a result of a substantial reorganisation of the initial twin by pronounced local lattice rotation and formation of boundaries, visible in the kernel average misorientation map in Figure 14h.

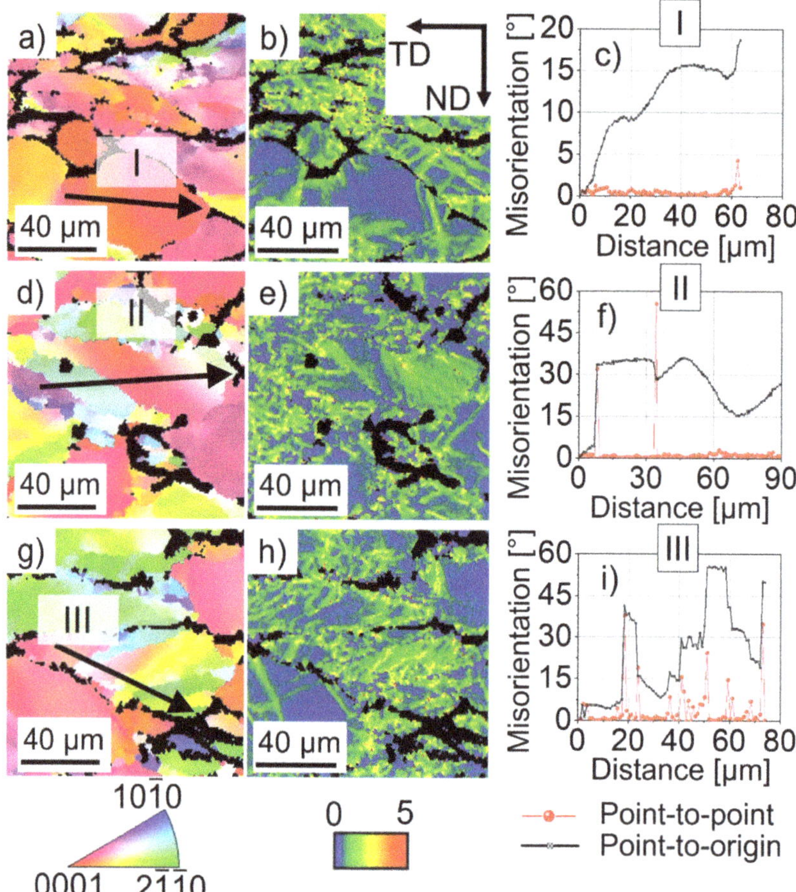

Figure 14. Details of the magnified areas A1 (shown in Figure 4), A2, and A3 (shown in Figure 5) for the (**a–c**) ZK40-CaO deformed at 250 °C and 0.001 s^{-1}; (**d–f**) ZK40-Gd deformed at 300 °C and 0.0001 s^{-1}; (**a**,**d**,**g**) inverse pole figure maps; (**b**,**e**,**h**) kernel average misorientation profile maps; (**c**,**f**,**i**) misorientation profiles I, II and III, respectively. Only α-Mg is indexed. Non-indexed regions are highly deformed regions or intermetallic compounds. ND and TD indicate the compression and transversal directions, respectively.

5. Summary and Conclusions

The deformation behaviour at 250 °C and 300 °C of the ZK40 alloy and modified ZK40 alloys with individual additions of CaO and Gd is investigated using in-situ synchrotron radiation diffraction and ex-situ characterisation techniques. The restoration mechanisms are derived from the interpretation of the microstructure features up to deformation of 0.3 of true strain. The following conclusion can be drawn:

- Additions of Ca and Gd increase the flow stress for the ZK40 alloys deformed at intermediated temperatures.
- Flow softening of ZK40-Gd at 300 °C occurs due to discontinuous dynamic recrystallisation and fragmentation of the intermetallic compounds.

- The fragmentation of intermetallic compounds seems to play a minor role in flow softening for the ZK40-CaO. Either only work hardening or a plateau in the flow stress occurs for this alloy up to the investigated strain.
- The controlling mechanisms that accommodate plastic deformation depend on the strain rate, temperature and strain, and are listed: misorientation spread within the α-Mg resulting from dislocation slip, the formation of low angle grain boundaries via dynamic recovery, twinning and dynamic recrystallisation.
- Dynamic recovery is the predominant restoration mechanism for the investigated alloys in all conditions.
- The formation of low angle grain boundaries occurs more homogeneously within the grain for ZK40 and ZK40-Gd. The Ca addition to the ZK40 seem to hinder the formation of subgrain boundaries during deformation. The formed boundaries are localised at the vicinity of original grain boundaries after 0.3 strain.
- Only the ZK40 deformed at 250 °C and 0.0001 s^{-1} shows continuous dynamic recrystallisation. This occurs via the formation of subgrains and the increase in boundary misorientation due to the progressive accumulation of dislocations via extensive cross-slip. It competes with discontinuous dynamic recrystallisation that occurs via bulging of subgrains and growth.
- Discontinuous dynamic recrystallisation is more pronounced at 300 °C compared to 250 °C and is present in the ZK40 and ZK40-Gd. Formation of an established subgrain structure that could lead to continuous dynamic recrystallisation is not found at 300 °C in any alloy.
- Ca addition to the ZK40 inhibits discontinuous dynamic recrystallisation.
- Gd addition to the ZK40 forms a semi-continuous network of intermetallic compounds that takes the load until its fragmentation, limiting the plastic deformation of the α-Mg, localising that at the grain boundaries. Thus, discontinuous dynamic recrystallisation is more limited for the ZK40-Gd compared to the ZK40.
- Tensile twins are rarely found in the microstructure after 0.3 strain. However, regions with similar morphology of deformation twins found throughout the microstructure for all conditions and alloys indicate twinning occurs at early stages of deformation followed by recovery and boundary formation within the twinned portion of the grains. New recrystallised grains formed in those regions are only found for the ZK40 alloy.

Author Contributions: Conceptualization, R.H.B., L.H.M.G., J.A.Á.D., D.T., C.L.M., N.H. and H.C.P.; Data curation, R.H.B.; Investigation, R.H.B., L.H.M.G., J.A.Á.D. and E.P.d.S.; Methodology, R.H.B., L.H.M.G., N.H. and H.C.P.; Resources, N.H. and H.C.P.; Software, R.H.B. and L.H.M.G.; Supervision, H.C.P.; Validation, R.H.B. and L.H.M.G.; Writing—original draft, R.H.B.; Writing—review & editing, L.H.M.G., J.A.Á.D., E.P.d.S., D.T., C.L.M., N.H. and H.C.P. All authors have read and agreed to the published version of the manuscript.

Funding: This research received no external funding.

Acknowledgments: The authors would like to thank the Brazilian Synchrotron Light Laboratory (LNLS) CNPEM/MCTIC for the use of the XTMS beamline. RHB acknowledges the CD-Laboratory for Design of High-Performance Alloys by Thermomechanical Processing with the support of the Christian Doppler Forschungsgesellschaft. L.H.M.G. acknowledges the financial Support of Coordenação de Aperfeiçoamento de Pessoal de Nível Superior—Brasil (CAPES) (Process No. 88882.180165/2018-01). J.A.Á.D. acknowledges CNPq for a postdoc scholarship (Grant No. 150215/2016-9). The authors sincerely acknowledge Maria Cecilia Poletti for fruitful and meaningful discussions. Open Access Funding by the Graz University of Technology.

Conflicts of Interest: The authors declare no conflict of interest.

Data Availability: The raw/processed data that support the findings of this study are available from the corresponding author, RHB, upon request.

References

1. Luo, A. Applications: Aerospace, automotive and other structural applications of magnesium. In *Fundamentals of Magnesium Alloy Metallurgy*; Pekguleryuz, M.O., Kainer, K.U., Arslan Kaya, A., Eds.; Woodhead Publishing: Oxford, UK, 2013; pp. 266–316.
2. Avedesian, M.; Baker, H. *ASM Specialty Handbook: Magnesium and Magnesium Alloys*; ASM International: Cleveland, OH, USA, 1999.
3. Yoo, M.H.; Morris, J.R.; Ho, K.M.; Agnew, S.R. Nonbasal deformation modes of HCP metals and alloys: Role of dislocation source and mobility. *Met. Mater. Trans. A* **2002**, *33*, 813–822. [CrossRef]
4. Ahlers, M. Stacking fault energy and mechanical properties. *Metall. Trans.* **1970**, *1*, 2415–2428. [CrossRef]
5. Sun, Z.; Wu, H.; Cao, J.; Yin, Z. Modeling of continuous dynamic recrystallization of Al-Zn-Cu-Mg alloy during hot deformation based on the internal-state-variable (ISV) method. *Int. J. Plast.* **2018**, *106*, 73–87. [CrossRef]
6. Humphreys, F.J.; Hatherly, M.; Humphreys, F.J.; Hatherly, M. Chapter 13—Hot Deformation and Dynamic Restoration. In *Recrystallization and Related Annealing Phenomena*; Elsevier: Oxford, UK, 2004; pp. 415–450. ISBN 9780080441641.
7. Takaki, T.; Yoshimoto, C.; Yamanaka, A.; Tomita, Y. Multiscale modeling of hot-working with dynamic recrystallization by coupling microstructure evolution and macroscopic mechanical behavior. *Int. J. Plast.* **2014**, *52*, 105–116. [CrossRef]
8. Li, H.; Wu, C.; Yang, H. Crystal plasticity modeling of the dynamic recrystallization of two-phase titanium alloys during isothermal processing. *Int. J. Plast.* **2013**, *51*, 271–291. [CrossRef]
9. Blum, W.; Zhu, Q.; Merkel, R.; McQueen, H. Geometric dynamic recrystallization in hot torsion of Al 5Mg 0.6Mn (AA5083). *Mater. Sci. Eng. A* **1996**, *205*, 23–30. [CrossRef]
10. Zhou, G.; Li, Z.; Li, D.; Peng, Y.; Zurob, H.S.; Wu, P. A polycrystal plasticity based discontinuous dynamic recrystallization simulation method and its application to copper. *Int. J. Plast.* **2017**, *91*, 48–76. [CrossRef]
11. Gourdet, S.; Montheillet, F. A model of continuous dynamic recrystallization. *Acta Mater.* **2003**, *51*, 2685–2699. [CrossRef]
12. Galiyev, A.; Kaibyshev, R.; Gottstein, G. Correlation of plastic deformation and dynamic recrystallization in magnesium alloy ZK60. *Acta Mater.* **2001**, *49*, 1199–1207. [CrossRef]
13. Couret, A.; Caillard, D. An in situ study of prismatic glide in magnesium—I. The rate controlling mechanism. *Acta Met.* **1985**, *33*, 1447–1454. [CrossRef]
14. Escaig, B. L'activation thermique des déviations sous faibles contraintes dans les structures hc et cc Par. *Phys. Status Solidi (b)* **1968**, *28*, 463–474. [CrossRef]
15. Legrand, P.B. Relations entre la structure électronique et la facilité de glissement dans les métaux hexagonaux compacts. *Philos. Mag. B* **1984**, *49*, 171–184. [CrossRef]
16. Wu, Y.-Z.; Yan, H.-G.; Zhu, S.-Q.; Chen, J.-H.; Liu, A.-M.; Liu, X.-L. Flow behavior and microstructure of ZK60 magnesium alloy compressed at high strain rate. *Trans. Nonferrous Met. Soc. China* **2014**, *24*, 930–939. [CrossRef]
17. Buzolin, R.; Tolnai, D.; Mendis, C.L.; Stark, A.; Schell, N.; Pinto, H.; Kainer, K.; Hort, N. In situ synchrotron radiation diffraction study of the role of Gd, Nd on the elevated temperature compression behavior of ZK40. *Mater. Sci. Eng. A* **2015**, *640*, 129–136. [CrossRef]
18. He, S.; Peng, L.; Zeng, X.; Ding, W.; Zhu, Y. Comparison of the microstructure and mechanical properties of a ZK60 alloy with and without 1.3 wt.% gadolinium addition. *Mater. Sci. Eng. A* **2006**, *433*, 175–181. [CrossRef]
19. Zhou, H.; Zhang, Z.; Liu, C.; Wang, Q. Effect of Nd and Y on the microstructure and mechanical properties of ZK60 alloy. *Mater. Sci. Eng. A* **2007**, *445*, 1–6. [CrossRef]
20. Ma, C.; Liu, M.; Wu, G.; Ding, W.; Zhu, Y. Tensile properties of extruded ZK60-RE alloys. *Mater. Sci. Eng. A* **2003**, *349*, 207–212. [CrossRef]
21. Homma, T.; Mendis, C.; Hono, K.; Kamado, S. Effect of Zr addition on the mechanical properties of as-extruded Mg–Zn–Ca–Zr alloys. *Mater. Sci. Eng. A* **2010**, *527*, 2356–2362. [CrossRef]
22. Li, Q.; Wang, Q.; Wang, Y.; Zeng, X.; Ding, W. Effect of Nd and Y addition on microstructure and mechanical properties of as-cast Mg–Zn–Zr alloy. *J. Alloy. Compd.* **2007**, *427*, 115–123. [CrossRef]
23. Kim, J.M.; Park, J.S. Microstructure and tensile properties of Mg–Zn–Gd casting alloys. *Int. J. Cast Met. Res.* **2011**, *24*, 127–130. [CrossRef]

24. Mordike, B. Creep-resistant magnesium alloys. *Mater. Sci. Eng. A* **2002**, *324*, 103–112. [CrossRef]
25. Singh, A.; Somekawa, H.; Mukai, T. Compressive strength and yield asymmetry in extruded Mg–Zn–Ho alloys containing quasicrystal phase. *Scr. Mater.* **2007**, *56*, 935–938. [CrossRef]
26. Stanford, N. Micro-alloying Mg with Y, Ce, Gd and La for texture modification—A comparative study. *Mater. Sci. Eng. A* **2010**, *527*, 2669–2677. [CrossRef]
27. Bohlen, J.; Nürnberg, M.R.; Senn, J.W.; Letzig, D.; Agnew, S.R. The texture and anisotropy of magnesium–zinc–rare earth alloy sheets. *Acta Mater.* **2007**, *55*, 2101–2112. [CrossRef]
28. Hoseini-Athar, M.; Mahmudi, R.; Babu, R.P.; Hedström, P. Effect of Zn addition on dynamic recrystallization behavior of Mg-2Gd alloy during high-temperature deformation. *J. Alloy. Compd.* **2019**, *806*, 1200–1206. [CrossRef]
29. Kwak, T.; Lim, H.; Kim, W. Effect of Ca and CaO on the microstructure and hot compressive deformation behavior of Mg–9.5Zn–2.0Y alloy. *Mater. Sci. Eng. A* **2015**, *648*, 146–156. [CrossRef]
30. Hofstetter, J.; Rüedi, S.; Baumgartner, I.; Kilian, H.; Mingler, B.; Povoden-Karadeniz, E.; Pogatscher, S.; Uggowitzer, P.; Löffler, J.F. Processing and microstructure–property relations of high-strength low-alloy (HSLA) Mg–Zn–Ca alloys. *Acta Mater.* **2015**, *98*, 423–432. [CrossRef]
31. Suzuki, A.; Saddock, N.; TerBush, J.R.; Powell, B.; Jones, J.; Pollock, T. Precipitation Strengthening of a Mg-Al-Ca–Based AXJ530 Die-cast Alloy. *Met. Mater. Trans. A* **2008**, *39*, 696–702. [CrossRef]
32. Luo, A.A. Recent magnesium alloy development for elevated temperature applications. *Int. Mater. Rev.* **2004**, *49*, 13–30. [CrossRef]
33. Zhang, B.; Wang, Y.; Geng, L.; Lu, C. Effects of calcium on texture and mechanical properties of hot-extruded Mg–Zn–Ca alloys. *Mater. Sci. Eng. A* **2012**, *539*, 56–60. [CrossRef]
34. Chino, Y.; Ueda, T.; Otomatsu, Y.; Sassa, K.; Huang, X.; Suzuki, K.; Mabuchi, M. Effects of Ca on Tensile Properties and Stretch Formability at Room Temperature in Mg-Zn and Mg-Al Alloys. *Mater. Trans.* **2011**, *52*, 1477–1482. [CrossRef]
35. Wiese, B.; Mendis, C.L.; Tolnai, D.; Stark, A.; Schell, N.; Reichel, H.; Brückner, R.; Kainer, K.U.; Hort, N. CaO dissolution during melting and solidification of a Mg 10 wt.% CaO alloy detected with in situ synchrotron radiation diffraction. *J. Alloy. Compd.* **2015**, *618*, 64–66. [CrossRef]
36. Buzolin, R.; Mendis, C.L.; Tolnai, D.; Stark, A.; Schell, N.; Pinto, H.; Kainer, K.; Hort, N. In situ synchrotron radiation diffraction investigation of the compression behaviour at 350 °C of ZK40 alloys with addition of CaO and Y. *Mater. Sci. Eng. A* **2016**, *664*, 2–9. [CrossRef]
37. Kwak, T.; Kim, W. Hot compression behavior of the 1 wt% calcium containing Mg–8Al–0.5Zn (AZ80) alloy fabricated using electromagnetic casting technology. *Mater. Sci. Eng. A* **2014**, *615*, 222–230. [CrossRef]
38. Canelo-Yubero, D.; Poletti, C.; Warchomicka, F.; Daniels, J.; Requena, G. Load partition and microstructural evolution during hot deformation of Ti-6Al-6V-2Sn matrix composites, and possible strengthening mechanisms. *J. Alloy. Compd.* **2018**, *764*, 937–946. [CrossRef]
39. Lonardelli, I.; Gey, N.; Wenk, H.-R.; Humbert, M.; Vogel, S.C.; Lutterotti, L. In situ observation of texture evolution during $\alpha \rightarrow \beta$ and $\beta \rightarrow \alpha$ phase transformations in titanium alloys investigated by neutron diffraction. *Acta Mater.* **2007**, *55*, 5718–5727. [CrossRef]
40. Ullrich, C.; Martin, S.; Schimpf, C.; Stark, A.; Schell, N.; Rafaja, D. Deformation Mechanisms in Metastable Austenitic TRIP/TWIP Steels under Compressive Load Studied by in situ Synchrotron Radiation Diffraction. *Adv. Eng. Mater.* **2019**, *21*, 1801101. [CrossRef]
41. Chi, Y.; Zhou, X.; Qiao, X.; Brokmeier, H.; Zheng, M. Tension-compression asymmetry of extruded Mg-Gd-Y-Zr alloy with a bimodal microstructure studied by in-situ synchrotron diffraction. *Mater. Des.* **2019**, *170*, 107705. [CrossRef]
42. Liss, K.-D.; Yan, K. Thermo-mechanical processing in a synchrotron beam. *Mater. Sci. Eng. A* **2010**, *528*, 11–27. [CrossRef]
43. Erdely, P.; Schmoelzer, T.; Schwaighofer, E.; Clemens, H.; Czermak, P.; Stark, A.; Liss, K.-D.; Mayer, S. In Situ Characterization Techniques Based on Synchrotron Radiation and Neutrons Applied for the Development of an Engineering Intermetallic Titanium Aluminide Alloy. *Metals* **2016**, *6*, 10. [CrossRef]
44. Elsayed, F.R.; Hort, N.; Ordorica, M.A.S.; Kainer, K.U. Magnesium Permanent Mold Castings Optimization. *Mater. Sci. Forum* **2011**, *690*, 65–68. [CrossRef]
45. Faria, G. Exploring Metallic Materials Behavior Through In Situ Crystallographic Studies by Synchrotron Radiation. Ph.D. Thesis, University of Campinas, Campinas, Brazil, 2014.

46. ImageJ. Available online: https://imagej.nih.gov/ij/ (accessed on 20 August 2019).
47. Liss, K.-D.; Schmoelzer, T.; Yan, K.; Reid, M.; Peel, M.; Dippenaar, R.; Clemens, H. In situstudy of dynamic recrystallization and hot deformation behavior of a multiphase titanium aluminide alloy. *J. Appl. Phys.* **2009**, *106*, 113526. [CrossRef]
48. Buzolin, R.; Mohedano, M.; Mendis, C.L.; Mingo, B.; Tolnai, D.; Blawert, C.; Kainer, K.; Pinto, H.; Hort, N. As cast microstructures on the mechanical and corrosion behaviour of ZK40 modified with Gd and Nd additions. *Mater. Sci. Eng. A* **2017**, *682*, 238–247. [CrossRef]
49. Buzolin, R.; Mohedano, M.; Mendis, C.L.; Mingo, B.; Tolnai, D.; Blawert, C.; Kainer, K.U.; Pinto, H.; Hort, N. Corrosion behaviour of as-cast ZK40 with CaO and Y additions. *Trans. Nonferrous Met. Soc. China* **2018**, *28*, 427–439. [CrossRef]
50. Hradilova, M.; Montheillet, F.; Fraczkiewicz, A.; Desrayaud, C.; Lejček, P. Effect of Ca-addition on dynamic recrystallization of Mg–Zn alloy during hot deformation. *Mater. Sci. Eng. A* **2013**, *580*, 217–226. [CrossRef]
51. Barnett, M.R. Quenched and Annealed Microstructures of Hot Worked Magnesium AZ31. *Mater. Trans.* **2003**, *44*, 571–577. [CrossRef]
52. Barnett, M.; Keshavarz, Z.; Beer, A.; Atwell, D. Influence of grain size on the compressive deformation of wrought Mg–3Al–1Zn. *Acta Mater.* **2004**, *52*, 5093–5103. [CrossRef]
53. Snir, Y.; Ben-Hamu, G.; Eliezer, D.; Abramov, E. Effect of compression deformation on the microstructure and corrosion behavior of magnesium alloys. *J. Alloy. Compd.* **2012**, *528*, 84–90. [CrossRef]
54. Dobroň, P.; Chmelík, F.; Yi, S.; Parfenenko, K.; Letzig, D.; Bohlen, J. Grain size effects on deformation twinning in an extruded magnesium alloy tested in compression. *Scr. Mater.* **2011**, *65*, 424–427. [CrossRef]

Publisher's Note: MDPI stays neutral with regard to jurisdictional claims in published maps and institutional affiliations.

Article

GTN Model-Based Material Parameters of AZ31 Magnesium Sheet at Various Temperatures by Means of SEM In-Situ Testing

Thorsten Henseler [1,*], Shmuel Osovski [2], Madlen Ullmann [1], Rudolf Kawalla [1] and Ulrich Prahl [1]

1 Institute of Metal Forming, Technische Universität Bergakademie Freiberg, 09599 Freiberg, Germany; madlen.ullmann@imf.tu-freiberg.de (M.U.); rudolf.kawalla@imf.tu-freiberg.de (R.K.); ulrich.prahl@imf.tu-freiberg.de (U.P.)

2 Faculty of Mechanical Engineering, Technion, 32000 Haifa, Israel; shmuliko@technion.ac.il

* Correspondence: thorsten.henseler@imf.tu-freiberg.de; Tel.: +49-3731-39-4179

Received: 1 September 2020; Accepted: 22 September 2020; Published: 23 September 2020 

Abstract: Magnesium alloys are primarily associated with complex forming mechanisms, which yield ductility at high temperatures. In sheet metal forming, high triaxiality stress states that favor the ductile damage mechanisms of void formation and growth are known to malleable metals. The formulation of coupled damage models has so far failed, due to the incomplete experimental determination of damage parameters for magnesium AZ31 thin sheet. A quantitative investigation was conducted to determine the ductile damage behavior of twin-roll cast, hot rolled, and annealed AZ31 thin sheet. Results on the mechanisms of void nucleation-, coalescence- and growth-rate were established at temperatures ranging from room temperature to 350 °C. In-situ tensile tests were carried out in a scanning electron microscope with three different specimen types: Simple tension specimens, notched specimens for high triaxiality stress state testing, and shear specimens. Through a comparative analysis of local strains measured by digital image correlation and local void volume fractions determined through post-mortem analysis of specimen cross-sections, GTN (Gurson–Tvergaard–Needleman) model-based material parameters were determined by experiment, representing a novel departure in the magnesium research landscape. The procedure developed in this context should also be transferable to other metals in the form of thin sheets.

Keywords: ductile damage; GTN model; magnesium; in-situ; deformation mechanisms

1. Introduction

The modeling of materials and the simulation of entire process chains are state-of-the-art tools in the development of components and their manufacturing processes. Requirements for high performance and low costs drive the demand for precise material models. Non-linear finite element (FE) programs with failure criteria or simple micromechanics-based models have long since they found their way into industrial developments, mostly because commercially available FE software has progressively implemented several damage models. Meanwhile, these include more complex coupled damage models and can further account for multi-axial and non-proportional loadings [1].

With their great lightweight potential, magnesium alloys are increasingly utilized in the automobile and aeronautical industries. Nevertheless, there are few guidelines for the determination of parameters of complex coupled damage models, and in the area of magnesium (Mg) thin sheet, there is a complete gap. Mg alloys have been of considerable interest for a long time from a scientific point of view, due to the properties associated with their hexagonal crystal lattice structure. At room temperature, the hexagonal close-packed (hcp) structure of Mg exhibits an inadequate amount of operational slip systems and substantial mechanical anisotropy, meaning that it is commonly regarded as a metal with poor ductility [2,3]. However, the ductility of Mg alloys with increasing temperature is mostly underestimated. Numerous studies confirm the strong increase in ductility Mg alloys exhibit at elevated temperatures, as well as the versatile interactions of individual deformation mechanisms in Mg, which may lead to superplasticity [4–7]. Samples made of Mg and its alloys tend to fail in a ductile manner at low temperatures. As such, the failure mechanisms consist of the nucleation of micro-sized voids, which grow and ultimately coalesce, leading to complete fracture. Nemcko and Wilkinson [8] made use of in-situ x-ray computed micro-tomography (XCT) during tensile tests to demonstrate the ductile failure process of pure Mg. In their work, void nucleation is predominantly evident at twin and grain boundaries, followed by a rapid growth stage leading to final fracture. The nucleation of voids at room temperature is generally understood to take place at grain boundaries, where second-phase particles are located, or due to twinning [9,10]. As what is becoming a popular method to study void development during deformation, XCT provides accurate results for statistical evaluation and overcomes drawbacks of two-dimensional post-mortem studies. In many studies on magnesium-aluminum-zinc alloys, void growth and coalescence are seen to result in apparent acceleration of the damage kinetics [11–14]. The process of void coalescence leads to an abrupt increase in the overall porosity, as the voids' growth is accelerated, due to their mutual interactions. Grain boundary sliding (GBS) is associated with complex void shapes [12]. The role of second-phase particles as major nucleation sites for voids is well established, although they do not influence the rate of void growth. Rashed et al. [13] find that with increasing strain, progressively smaller particles induce voids. Furthermore, it remains possible that the material could exhibit pre-existing hydrogen micro-voids resulting from the precipitation of hydrogen during solidification after the casting process. These have a considerable influence on increases in void volume fraction, as they exhibit the same growth rates as strain-induced voids. Compared to voids on grain boundaries, pre-existing grain-interior micro-voids grow quite slowly [14]. Kondori and Benzerga [15] draw attention to the fact that the strong basal texture of rolled magnesium shows greater tolerance to ductile damage accumulation in notched specimens than in smooth ones. Here, a transition from twinning-induced fracture under uniaxial tension to void coalescence fracture under triaxial loading was evident.

In this study, the ductile damage behavior of twin-roll cast, hot rolled, and annealed AZ31 sheet is characterized by in-situ tensile testing, as well as a two-dimensional post-mortem cross-section analysis in a scanning electron microscope (SEM). Then, Gurson–Tvergaard–Needleman (GTN) model-based ductile fracture parameters are proposed as a function of temperatures ranging from room temperature (RT) to 350 °C. These investigations get to the bottom of the current gap in the state-of-the-art to establish a characterization method that allows the experimental determination of damage parameters in thin sheet metal. The GTN model is based on the approach by Gurson [16] and is the most widely used continuum mechanics-based model. In the absence of porosity, the Gurson flow potential reduces to the von Mises potential, while as porosity grows, the Gurson flow potential contracts to express the loss of load-bearing capacity as a function of the evolving void volume fraction. From Tvergaard, Chu, and Needleman [17–19], an extension of the Gurson model has proven to offer more accuracy. The resulting flow potential is defined as:

$$\Phi_{\mathrm{GTN}} = \left(\frac{\sigma_{\mathrm{v}}}{\sigma_{\mathrm{y}}}\right)^2 + 2q_1 f^* \cosh\left(\frac{3}{2} q_2 \frac{\sigma_{\mathrm{H}}}{\sigma_{\mathrm{y}}}\right) - \left(1 + q_3 f^{*2}\right) \tag{1}$$

where σ_{y} is the yield stress of the matrix material, σ_{v} is the von Mises equivalent stress and σ_{H} is the mean stress. Tvergaard, Chu, and Needleman replaced the void volume fraction f by a modified damage parameter f^* to reflect void coalescence processes after a critical void volume fraction f_{c} is reached:

$$f^* = \begin{cases} f & , \text{ if } f < f_{\mathrm{c}} \\ f_{\mathrm{c}} + \frac{f_{\mathrm{u}}^* - f_{\mathrm{c}}}{f_{\mathrm{f}} - f_{\mathrm{c}}}(f - f_{\mathrm{c}}) & , \text{ if } f \geq f_{\mathrm{c}} \end{cases} \tag{2}$$

where f_f is the critical void volume fraction at fracture and $f_u^* = q_1 + \sqrt{q_1^2 - q_3}/q_3$. By adding the material-dependent parameters q_1, q_2 and q_3, a better fit of numerical results can be achieved [17–19]. The model assumes a homogeneous distribution of spherical voids in an isotropic material. f follows the evolution equation $\dot{f} = \dot{f}_{\text{nucleation}} + \dot{f}_{\text{growth}} = \dot{f}_{\text{nuc}} + \dot{f}_{\text{gr}}$, with:

$$\dot{f}_{\text{nuc}} = \frac{f_N}{S_N\sqrt{2\pi}} \cdot \exp\left[-\frac{1}{2}\left(\frac{\overline{\varepsilon}^p - \varepsilon_N}{S_N}\right)^2\right] \cdot \dot{\overline{\varepsilon}}^p \tag{3}$$

$$\dot{f}_{\text{gr}} = (1 - f) \cdot \text{trace}\left(\dot{\varepsilon}^p\right) \tag{4}$$

where f_N is determined so that the void volume fraction nucleated is consistent with the volume fraction of void-inducing second-phase particles. Furthermore, Tvergaard, Chu, and Needleman assume a normal distribution of void nucleation at second-phase particles with ε_N being the mean plastic strain at maximum nucleation and S_N the standard deviation.

Publications of successful continuum damage modeling applications in the forming of the AZ31 sheet are quite scarce [20–22]. In fact, this study represents a novel departure in the Mg research landscape on experimentally determined GTN model-based damage parameters. The authors investigate the adaptation of the model to use with a hexagonal metal.

2. Materials and Experimental Procedure

For the present work, twin-roll cast, hot rolled, and annealed 1.0 mm-thick AZ31 magnesium sheet was manufactured by the Institute of Metal Forming (Technische Universität Bergakademie Freiberg, 09599 Freiberg, Germany). The chemical composition of the investigated alloy is presented in Table 1. A description of the manufacturing process and the expected final properties of the test material can be found in numerous publications [23–28].

Table 1. Nominal chemical composition of the investigated AZ31 magnesium sheet (wt.%).

Al	Zn	Mn	Zr	Cu	Si	Fe	Sn	Ca	Mg
2.99	0.969	0.377	0.002	0.001	0.017	0.003	<0.005	<0.001	bal.

First, rough outlines of the tensile specimens were cut from the sheet via waterjet cutting. Subsequently, the stressed parts of the specimen were fine milled to precise dimensions, as shown in Figure 1b–d. The work sequence for the experimental procedure was that of initial surface preparation, then tensile testing, and finally, the reconstruction of the void volume fraction from the sample cross-sections. For the local strain analysis and the observation of near-surface damage phenomena, in-situ (during the experiment) images were taken by means of a MIRA3 scanning electron microscope (SEM) from Tescan (at the Technion, 32000 Haifa, in the lab for Computational and Experimental Micromechanics of Materials).

The tensile tests were carried out with an in-situ tension-compression module from *Kammrath and Weiss* with a traverse velocity of 5 µm/s (see Figure 1a). The heating stage was located beneath the specimen. The loading direction was parallel to the rolling direction at 20 °C, 150 °C, 250 °C, and 350 °C. The tensile tests were preceded by heating of the specimen at a rate of 1 K/s and a holding time of 20 s.

The surface preparation of the samples was carried out mechanically and chemically. In order to avoid excessive removal of the material on the surface, only three preparation steps were implemented as preparation for the EBSD (electron backscatter diffraction) analysis, as well as for the in-situ testing. The first step involved a grinding stage (with the addition of ethanol) on P4000 grade SiC abrasive paper. The samples were then polished with a SiO_2 + iron oxide suspension (particle size 0.06 µm). The last step involved immersing the samples in Nital (95 mL of ethanol and 5 mL of concentrated nitric acid) for not more than 5 s. The Nital treatment removes the slightly deformed polished surface in order to ensure superior EBSD results. The EBSD analysis (15 kV acceleration voltage, 0.6 µm step size) was carried out using *Aztec* software from *Oxford Instruments* and subsequent evaluation through an open-source MATLAB toolbox called MTEX (for the calculation of the orientation distribution function and pole figures from the EBSD data) [29]. The correct half-width of 7.5° and bandwidth of 33 was determined by means of the optimal kernel function algorithm calcKernel and the RuleOfThumb method for orientation distribution function (ODF) estimation available in *MTEX*.

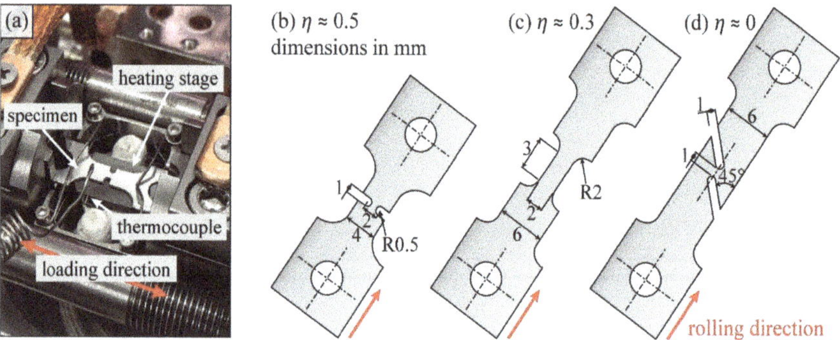

Figure 1. (**a**) Experimental setup for in-situ tensile tests inside the SEM, as well as specimen geometries in mm and their theoretical stress triaxiality η for stress states of (**b**) high triaxiality, (**c**) simple tension, and (**d**) simple shear.

To determine the void volume fraction in a cross-section parallel to the sheet plane, torn samples were ground to half their thickness. Here, the sample holder AccuStop from Struers was used to ensure good planarity. The surface preparation was carried out in the same way as described above. High-resolution SEM images where used to determine the area fraction (see Figure 2), as well as the size distribution of voids and second-phase particles.

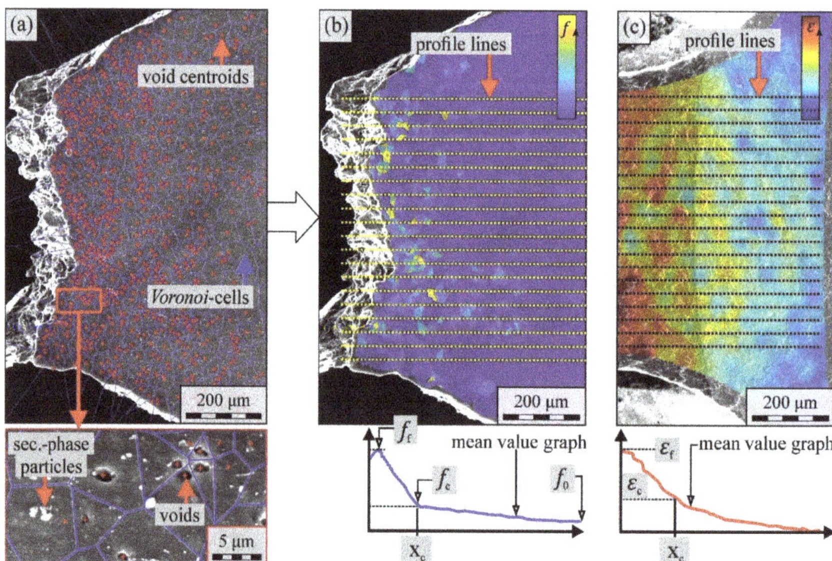

Figure 2. Voronoi decomposition method to calculate the local void area fraction by (**a**) partitioning the cross-section images and (**b**) relating the individual region areas to the void area contained therein; (**c**) example of digital image correlation (DIC) strain analysis carried out using images of the specimen surface.

Since the present study was carried out over a wide range of temperatures and stress triaxialities, which manifested in rather large variations in local void densities, the authors opted for a method known as *Voronoi* decomposition [30]. The borders of the Voronoi cells were positioned halfway between the two neighboring void centroids and formed the reference area in which the local void area fraction was determined, as seen in Figure 2a,b. The void area fraction f was determined on the specimen cross-section parallel to the sheet plane. Thus, the simplification was made that the area fraction of visible voids corresponded to the void volume fraction, while the area of the second-phase particles corresponded to the particle volume fraction.

The strain analysis was carried out from the discrete SEM images by means of digital image correlation (DIC, see Figure 2c). The natural surface structure provided the necessary contrast in the present work. The MATLAB tool *Ncorr v1.2* used is an open-source, two-dimensional DIC program whose algorithms are described in Reference [31]. A total of 20 profile lines were placed along the loading direction on the local void area fraction data, as well as strain analysis. As demonstrated in Figure 2 mean value graphs were then used to correlate the mean local void area fraction with local strains.

3. Results and Discussion

3.1. Materials Characterization

The initial microstructure of the investigated AZ31 magnesium sheet was characterized by globular grains, with an average equivalent grain diameter of 6 μm determined with through-thickness EBSD data (2° grain tolerance angle and 2 μm minimum grain size, see Appendix A, Figure A1). As demonstrated in Figure 3a larger grains (which are predominantly located at the sheet upper and lower surface) have a size of up to 50 μm, and reveal a minor bimodal grain size distribution. The grain size distribution does not vary between cross-sections 0° or 90° to rolling direction (RD).

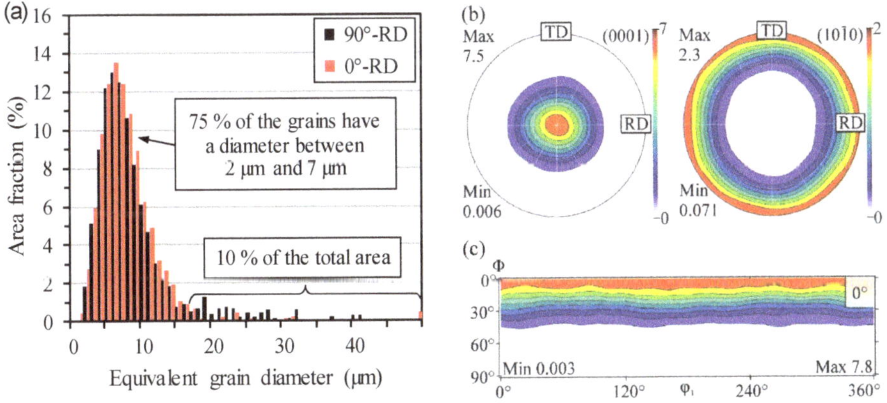

Figure 3. The initial microstructure of the AZ31 magnesium sheet studied, characterized by (**a**) the grain size distribution with reference to the respective area fraction, (**b**) pole figures with reference to the basal and prismatic planes, as well as (**c**) the orientation distribution function (ODF) section at $\varphi_2 = 0°$.

The presented pole figures in Figure 3b confirm a strong basal texture with the (0001) basal poles showing a high intensity normal to the RD-TD (rolling direction-transverse direction) plane. This is a known phenomenon in rolled Mg sheets, with the basal planes reoriented parallel to the horizontal material flow as basal dislocation slip is activated in the rolling direction [24,32]. Furthermore, the pole figure for the (10-10) prismatic planes shows a random distribution in sheet plane, revealing a typical fiber texture. The calculated ODF sections at $\varphi_2 = 0°$, shown in Figure 3c, confirm the fiber texture.

The intragranular misorientation angles from the EBSD data was determined to have a median value of 0.3 (map shown in Appendix A, Figure A1). The authors can, therefore, presume that the microstructure was in a completely recrystallized and recovered state prior to tensile testing.

3.2. Deformation Mechanisms

In the following, in-situ SEM images from the specimen surface during loading are discussed. Here, the authors acknowledge that in the present investigations, the observed phenomena at the surface might not quite correspond to the events inside the sample volume. Surface observation represents a shear stress-free interface and offers the opportunity to gain new impressions about material mechanisms. In the unloaded condition, the specimen surface has a dark grey appearance with uniformly distributed second-phase particles in bright contrast ranging from 0.5 to 5 μm in size (see Appendix A, Figure A2). These particles have a composition that matches the alloying elements according to the γ-phase $Mg_{17}Al_{12}$ (larger particles) or Al-Mn-containing intermetallic compounds (smaller particles at grain boundaries) [33,34]. Due to the short etching time and the selected contrast conditions, no grain boundaries could be identified on the surface. As illustrated in Figure 4a–c for a simple tension specimen at 150 °C, grain distortion, grain reorientation through GBS, and slip traces from dislocation glide can clearly be identified as soon as loading is applied. By tracking individual particles (red circles in Figure 4a–c) the evolution of the surrounding grains could be observed as a function of applied strain. When comparing the sample surfaces in Appendix A, Figure A2, it can be seen that a surface-relief is induced with rising strains attributed to GBS (also see Reference [35]). Features in the images taken from high triaxiality and simple tension specimens did not differ. As indicated in Figure 4b, grain distortion took place along the loading direction, while the slip traces predominantly became visible at an angle of 45°. Especially at RT and 150 °C, these observations could be made quite clearly. In a similar study by Xia et al. [36] with Mg-AZ31 at RT, particular attention was paid to inhomogeneous strain fields, which cause the distortion of grains. In Reference [37], shear bands in AZ31, as well as in pure magnesium [38], are seen to be activated early and incur heavy local strains. As soon as GBS occurs, the grain boundaries, which are dimensionally like the EBSD initial state analysis, become clearly visible.

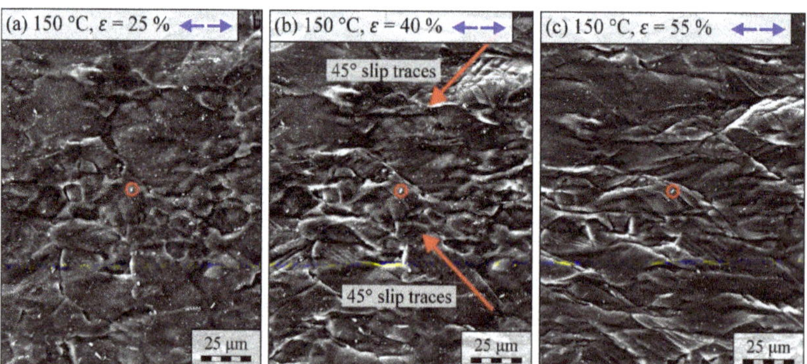

Figure 4. Example of a 150 °C simple tension specimen (loading direction indicated by blue arrows) with a tracked particle indicated by red circles and the orientation of slip traces indicated by red arrows; elongation ε indicated with reference to the initial gauge length of 3 mm being (**a**) 25%, (**b**) 40% and (**c**) 55%.

As can be expected, the simple shear specimen changed the stress mode, inducing an entirely different pattern at the specimen surface than in the tension specimens (see Figure 5a–c). Here, the visible slip traces appeared parallel to the loading direction, while the grains were distorted and rotated 45° away from it. This shows that the specimen geometry had, in fact, induced shear deformation along the loading direction. Nevertheless, the authors could also observe that with increasing deformation of the shear specimen, the grains and slip traces rotated back. At high strains (possible at 250 °C and 350 °C, though not shown Figure 5), this rotation ended in the grains aligning parallel to the loading direction and leaving a similar appearance as simple tension samples (comparable to Figure 4). Twinning could not be identified clearly in the present study, although the results of Reference [39] suggest that the parallel lines could be a sign thereof. If the orientation of slip traces within a grain changes suddenly, Reference [40] determined that this has to do with a local change in the active slip system. According to Reference [41], the change in slip systems during increasing local strains is related to increasing levels of connectivity between favorably oriented grains. It is stated in Reference [38] that twinning observed during in-situ examinations on the grain-scale does not follow the orientation of slip traces and is clearly identifiable. The perpendicular orientation of the loading direction towards the sharp basal texture points out, however, that the Schmid-factor of prismatic and pyramidal slip must be highest. In the present study, the occurrence of both twinning and basal slip is unlikely, especially at elevated temperatures.

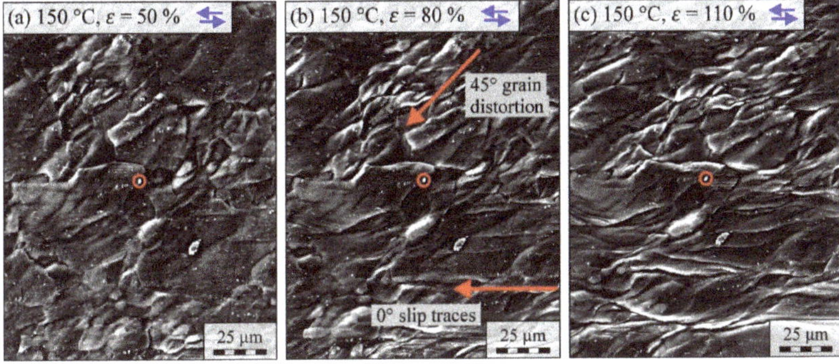

Figure 5. Example of a 150 °C simple shear specimen (loading direction indicated by blue arrows) with a tracked particle indicated by red circles and the orientation of slip traces indicated by red arrows; elongation ε indicated with reference to the initial gauge length of 1 mm being (**a**) 50%, (**b**) 80% and (**c**) 110%.

The occurrence of both the phenomena of GBS, as well as of slip trace formation, exhibited a clear temperature-dependence (see Figure 6). As soon as loading was applied and as the first visible mechanism of deformation, GBS began to change the surface topography. The temperature-dependence can be observed in Figure 6a–d, where rising temperatures increased the activity, and thus, the contribution of GBS to deformation. As a result, the rising temperature decreased the appearance of slip traces (Figure 6e–h). The authors concluded that due to softening mechanisms, the dislocation density did not increase enough to make slip traces visible. The GBS observed at room temperature was considered to be related to locally inhomogeneous deformation, while pure GBS without shear band formation was the cause of surface undulation at temperatures above 250 °C, as stated by Reference [35]. As the critical resolved shear stresses of non-basal slip systems fall with increasing temperature, the probability of slip traces arising from a single slip mode per grain decreases greatly. According to Reference [41], the influence of grain sizes of neighboring grains has a great influence on the active deformation mechanism, which was why differences between single grains could be observed in this study as well. The authors of the present study further assume that GBS is generally more active at the surface, as these grains have more freedom of movement than grains in the bodies of the samples.

Lastly, a phenomenon that the authors presume to be a sort of dynamic recrystallization became visible in the later stages of 350 °C testing only. While extremely elongated grains and highly roughened topology characterized the surface of the 250 °C specimen (Figure 7a), the surface of the 350 °C specimen (Figure 7b) exhibited relatively equiaxial grains. The low distortion of these equiaxial grains at 350 °C suggested that they were not significantly involved in the deformation process. Leaving the surface open-pored, numerous smaller grains appeared increasingly in the interstices of these grains. As seen in Figure 7c, these smaller grains had angular—and no smooth—edges. These specific in-situ surface observations were unique to this study. Therefore, an unmistakable relation to the well-known mechanism of dynamic recrystallization with reference to supplementary studies could not be made.

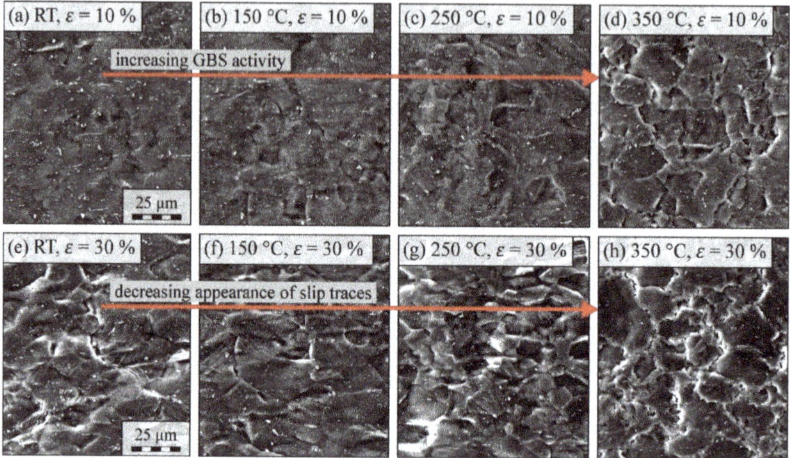

Figure 6. Increasing grain boundary sliding (GBS) activity observed for simple tension specimens at (**a**) RT, (**b**) 150 °C, (**c**) 250 °C, and (**d**) 350 °C, and strains equal to 10%; varying slip trace appearance observed for simple tension specimens at (**e**) RT, (**f**) 150 °C, (**g**) 250 °C, and (**h**) 350 °C, and strains equal to 30%.

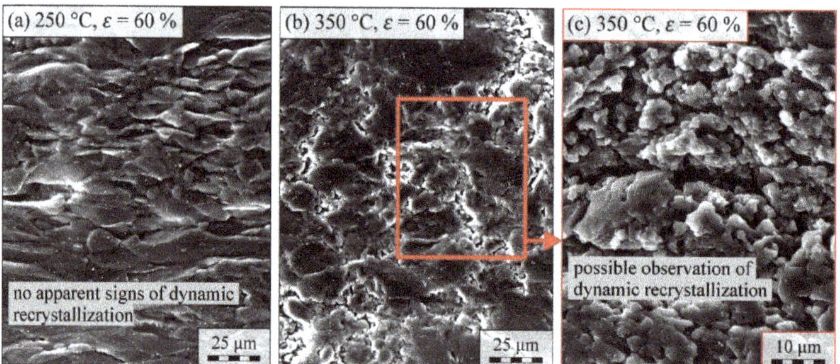

Figure 7. Comparison of the (**a**) 250 °C specimen to the (**b**,**c**) 350 °C in-situ simple tension test specimen where dynamic recrystallization reduced grain elongation and formed new grains along existing boundaries.

3.3. Damage Mechanisms

3.3.1. Initial Void Volume Fraction

The initial void volume fraction was determined on a non-loaded sample, with $f_0 = 0.0004 \pm 0.0001$, by surface area fractions of matrix versus voids visible in the sample cross-section. It is reasonable to assume that the voids in f_0 originated from the TRC manufacturing process. This involved a very small proportion of existing micro-voids that result from the precipitation of hydrogen during solidification [14]. These voids were not eliminated in the subsequent hot rolling process. Based on a numerical evaluation, i.e., without experimental investigations, Wang et al. [20] and Zhao et al. [21]

apply an initial void volume fraction of 0.001 for AZ31, which is twice as high. However, f_0 represents a property that is strongly dependent on the manufacturing process—e.g., casting or forming—and that should be determined separately for each material investigated.

3.3.2. Void Nucleation

The GTN model considers the dynamics of newly nucleated voids as a function of void-inducing particles. In the present work, this corresponds to the surface area fraction of the second-phase particles f_N. The assumption that the totality of these particles could initiate the nucleation of voids corresponds to the definition in the primary literature [17,19]. Minus the initial void volume fraction in the non-loaded cross-section of a sample, f_N was determined to be equal to 0.011 ± 0.003. Analogously to the initial void volume fraction, this value is strongly dependent on the microstructure (precipitation, phases, inclusions, etc.) and should not simply be taken from the literature. The mean strain for void nucleation ε_N means that half of all voids that could be initiated by second-phase particles were formed at $\overline{\varepsilon}^p = \varepsilon_N$. The void nucleation rate $\dot{f}_{nuc}$ is subject to a *Gaussian* normal distribution with the standard deviation S_N, see Equation (3). In the present work, ε_N and S_N were evaluated numerically, since they can only be obtained indirectly from the explicit representation of the void-development function using experimental data (see Figure 2, bottom graphs).

3.3.3. Void Coalescence and Failure

The void volume fraction at the onset of void coalescence f_c was determined by means of the mean value of several line sections in the sample cross-section parallel to the direction of loading. Each line section indicated the course of $f(x)$ as a function of the location x. As soon as void coalescence occurred, the increase in the curve changed and f_c could be read directly from it (see Figure 2). The void volume fraction at the moment of failure f_f was determined analogously, as the average value of the maxima of $f(x)$. The locations x for f_c and f_f are of crucial importance for the calibration of the GTN model because they make the correlation to the local strain possible. After the scales of the SEM images and the DIC analysis were adjusted, it was possible to read ε_c and ε_f directly. f_c and f_f were equally temperature-dependent (Figure 8a,b) and followed an almost linear relationship with increasing temperature. At RT, f_c and f_f assumed the lowest values for both sample types. The results show the largest void volume fraction at 350 °C.

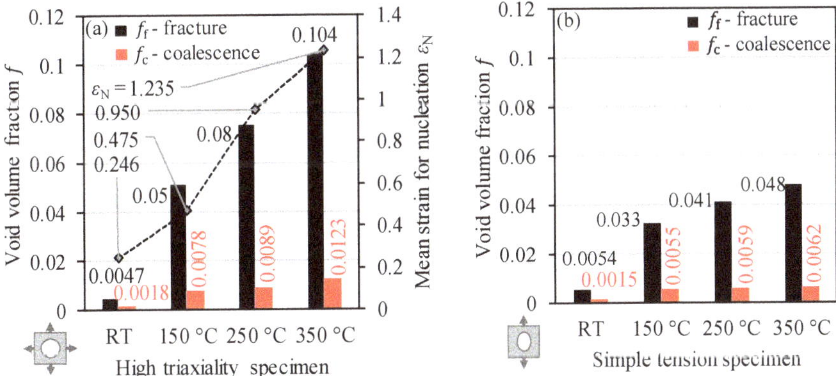

Figure 8. Gurson–Tvergaard–Needleman (GTN) model-based material parameters for critical void volume fractions for fracture and the onset of void coalescence, as well as mean strains for nucleation at RT, 150 °C, 250 °C, and 350 °C for (**a**) a high-stress triaxiality specimen and (**b**) a simple tension specimen.

During the determination of the void volume fraction, it was qualitatively evident that the voids became larger and larger from RT to 350 °C. This feature was also apparent in the analysis of the fracture surfaces. In addition, it was found that f_c and f_f in the notched tensile specimens were about twice as high as in the uniaxial tensile test. The high triaxiality of notched tensile specimens leads to rapid void growth, which explains the higher absolute values [15]. Ray and Wilkinson [9] indicate an f_f value at RT of less than 0.005 for AZ31, which is significantly lower than the values found in the present work. Possible reasons for this are that the tensile specimens in Reference [9] had a very large notch radius, the computer tomography used only detected voids with a size of 5 μm or more, and that the tests were stopped. This means that at the time of analysis, the samples were just about to fail; on the contrary, in the present analysis, the specimens have reached failure elongation.

3.3.4. Role of Second Phase Particles

The evolution equation of the void volume fraction in the GTN model functions independently of the actual mechanism of void nucleation. The model cannot distinguish whether voids have formed in the vicinity of particles, at grain-boundary triple points, or at twin boundaries without explicitly introducing the nucleation microstructural feature into the model, as was done in References [42,43]. Since the second-phase particles were homogeneously distributed and did not form accumulations, premature failure, as in the investigations of Wang et al. [44], could not be assumed in this respect. Barnett [45] also describes a direct relationship between twin formation and void nucleation. Thus, twins initiate void nucleation within a grain, which then grow rapidly up to the grain boundaries. This includes, in particular, double twins—which, according to Ulacia et al. [46], are more likely to occur at high strain rates. The present investigations indicate that void nucleation must have taken place not only through second-phase particles, but also at grain boundaries. When considering single-phase metals, Bieler et al. [47] show that void nucleation is mainly caused by damaging alternating effects of dislocations at grain boundaries. Against this background, the representative accuracy of the damage model could be optimized by adding further terms for the respective void nucleation mechanisms, such as in Reference [48]. However, this would also mean that the dynamics of the respective void nucleation mechanisms could be quantified.

In order to clarify the void nucleation mechanism, a distinction is made in the following between particle breakage and decohesion from the matrix. The matrix material appears greyish, voids are significantly darker or black, and particles are white. The mechanism of void nucleation is strongly dependent on the material properties, such as particle strength, size, and shape, as well as the strain hardening behavior of the matrix. According to Benzerga and Leblond [49], the decohesion process occurs in soft matrix materials, while particle breakage occurs in hard matrix materials. In the present work, it was observed that a series of large fragmented second-phase particles (>5 μm) were present in areas of low strain (Figure 9a). In areas of high plastic strain, especially near the fracture surface, the particle size rarely exceeded one micrometer (see Figure 9b). Spherical voids without direct contact with a particle may have been in the matrix already and were the result of the initial void volume fraction.

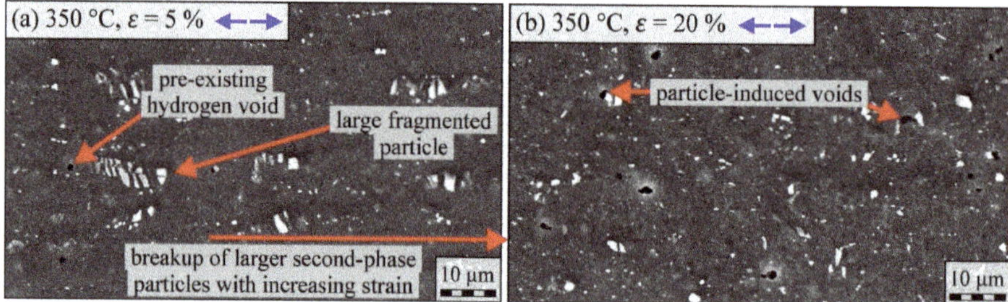

Figure 9. Exemplary SEM images of the sample cross-section of a sample tested at 350 °C (**a**) slightly further from the fracture surface with low strains and (**b**) near the fracture surface with correspondingly higher local strain.

Large particles were broken by the deformation process and then acted as nucleation sites for void formation. The fact that the mechanisms of both particle breakage and decohesion from the matrix were active was confirmed by SEM images. As can be observed in Figure 10a, the breakage of the particle (white) led to the nucleation of the void. From a particle size of approximately 1 μm upwards, however, the mechanism of decohesion was dominant (Figure 10b). In the case of the investigated material, which mechanism of void nucleation was dominant depended on the particle size. It was also found during the investigations that some voids contained several particles. This is possibly an indication that voids that were once formed individually had coalesced at high strain levels.

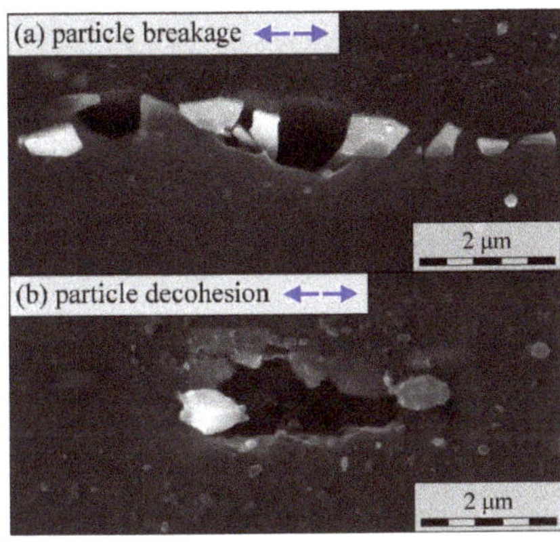

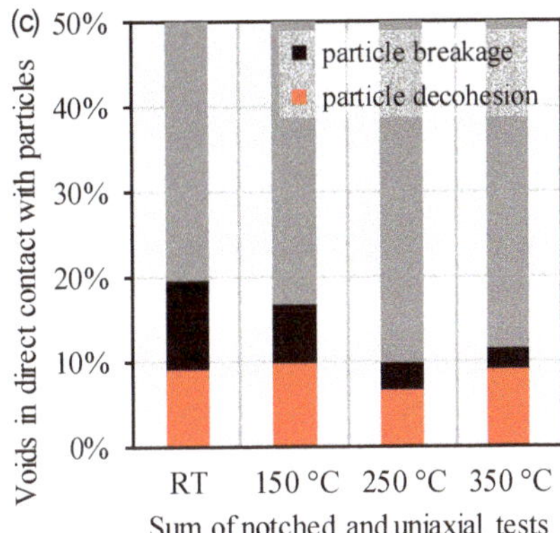

Figure 10. SEM images of a particle breakage (**a**) and the decohesion of a particle from the matrix (**b**), which led to void nucleation in the notched specimen at 200 °C; Statistics on the analyzed voids that were in direct contact with particles after the test (**c**).

In addition to the analysis of the void volume fraction, another quantified factor was how many of the voids were still in direct contact with particles after the test. In each of the cross-sections for the statistics presented in Figure 10c, an average of 2000 to 3000 voids were evaluated. Lhuissier et al. [11] find that 30% of the voids observed in AZ31 hot tensile tests were in direct contact with particles. In the present study, this value lay between 10 and 20% depending on the temperature.

It may be assumed that the actual proportion of voids that were in direct contact with particles during deformation was higher, since the bond between particles and the matrix is significantly reduced during void growth. In addition, the mechanical grinding and subsequent polishing of the cross-sections removed many particles, which distorted these statistics. Furthermore, existing hydrogen micro-cavities, which do not necessarily contain particles, cannot be distinguished after the test [14]. The cavities do not necessarily have to be spherical, since the alloy under investigation exhibits high GBS activity at higher temperatures, and thus, develops complex cavity shapes [12]. Nemcko et al. [50] also show that the anisotropic material behavior of the magnesium crystal can lead to the nucleation of irregularly shaped voids.

With increasing temperature, the incidence of voids with broken particles decreases because the strength of the matrix decreases, and voids tend to form through decohesion from the matrix. The fact that at higher temperatures, there are fewer voids in contact with particles may be attributable to voids becoming significantly larger than the particles that initiate them. At low temperatures, more voids are formed, but they do not become quite as large, due to the low degree of local strain. Many particles, thus, remain trapped in the void. The distance between voids is a decisive factor for the coalescence phase, and thus, for the ultimate failure of the sample [51]. Because void nucleation and cavity growth occur over large strain ranges, many voids of different sizes are created. In addition, large voids influence the growth behavior of smaller voids in their environment [52]. This could be confirmed for the investigated material, and in the vicinity of the fracture surface.

3.3.5. Calibration of Growth Parameters and Normal Distribution

The GTN parameters q_1, also known as the growth parameters, are usually specified in the literature as $q_1 = 1.5$, $q_2 = 1$, and $q_3 = {q_1}^2 = 2.25$ in accordance with the recommendations of Tvergaard and Needleman [17]. They influence the shape and size of the yield surface and have a direct influence on the void growth rate $\dot{f}_{gr}$ through the flow rule. Figure 11a–d shows how the void volume fraction behaved in the numerical simulation as a function of the effective plastic strain $\bar{\varepsilon}^p$ for the model parameters developed in the present study. An explicit formulation of the void growth rate from Prahl et al. [53,54] was used for this representation, integrating the equations for a single material point. This was followed by the application of the associated flow rule (the equation of the evolution of $\bar{\varepsilon}^p$) in place of the trace of the strain rate tensor $\dot{\varepsilon}^p$ in

$$\dot{f}_{gr} = (1 - f)\cdot \mathrm{trace}\left(\dot{\varepsilon}^p\right) = (1 - f)\dot{\Lambda}\frac{\partial \Phi_{GTN}}{\partial \sigma_H}. \tag{5}$$

Under the assumption that flow occurs and that $\sigma_v = \sigma_y$ stands without feedback from f, the specific void volume fraction f_{gr} can be expressed as a function of the effective plastic strain $\bar{\varepsilon}^p$ [53]:

$$f_{gr} = f_0 \cdot \exp\left(\frac{3}{2}q_1 q_2 \sinh\left(\frac{3}{2}q_2\eta\right)\bar{\varepsilon}^p\right). \tag{6}$$

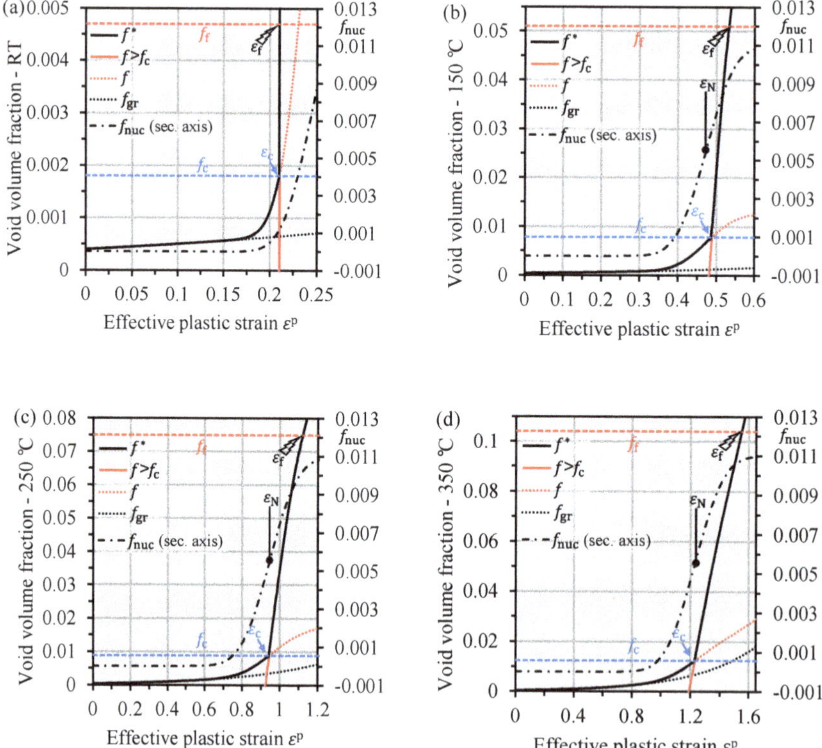

Figure 11. Void volume fractions after calibration of the growth parameters for the temperatures RT (**a**), 150 °C (**b**), 250 °C (**c**), and 350 °C (**d**) to fit experimental data taken according to Figure 2 from notched samples.

It was evident that the volume fraction of growing voids f_{gr} was dependent on the initial void volume fraction f_0 and the triaxiality factor η, as well as q_1 and q_2. In the following a triaxiality factor $\eta = 0.5$ is assumed. Since the GTN model also represents void nucleation, due to second-phase particles, the void growth rate $\dot{f}$ is extended by the void volume fraction of new voids f_{nuc}. This can be clearly seen in Figure 11, where $f = f_{nuc} + f_{gr}$. The sigmoidal curve of f_{nuc} (secondary axes) starts at zero and reaches its maximum at $f_{nuc} = f_N = 0.0011$. Viewed numerically, the newly formed voids are added to the initial void density as a function of the strain $\bar{\varepsilon}^p$, and then increase at the same rate. In the GTN model, ε_N is the parameter that indicates from which effective plastic strain half of all of the particles have led to the nucleation of a void. Its change results in the horizontal displacement of f_{nuc}. S_N is the standard deviation and indicates the slope in the sigmoidal curve.

f^* represents the partially continuous function which, in the flow potential, leads to a reduction of the flow area along the hydrostatic axis. According to Equation (2), a case distinction is made depending on f_c in order to determine the course of f^* in association with the active damage mechanisms. Therefore, the course of f^* first runs along f (nucleation + growth) and from f_c along "$f > f_c$" (nucleation + growth + coalescence). It is clear that the rate of f^* increases from f_c. At this point, the function is not continuous and can lead to a bend in the course of f^*. As soon as f^* is equal to f_f, the critically damaged elements are eliminated in the FEM simulation. It can be seen from the range of values of the horizontal axes in Figure 11 that the local strains varied greatly depending on the temperature. The rate of increase of the void volume fraction is approximately 0.02 at RT and 0.06–0.08 at temperatures ranging between 150 °C to 350 °C and a constant triaxiality of 0.5.

In the present work, the parameters q_1, q_2, ε_N and S_N, which could not be measured directly in the experiment, were numerically calibrated. Through the direct correlation of f_c and f_f with ε_c and ε_f (cf. Figure 11), clear boundary parameters could be established:

$$f^*(\bar{\varepsilon}^p = 0) = f_0 \tag{7}$$

$$f^*(\bar{\varepsilon}^p = \varepsilon_c) = f_c \tag{8}$$

$$f^*(\bar{\varepsilon}^p = \varepsilon_f) = f_f. \tag{9}$$

The calibration was performed under the assumption that constant stresses prevailed with a triaxiality factor of $\eta = 0.5$. This is closest to the mean stress state of the notched tensile specimens used. Due to the quadratic cross-section here, the generally assumed triaxiality factor of 0.6 from round notched samples cannot be assumed.

The regression analysis performed using the Solver add-in program in Microsoft Excel showed that it was possible to calibrate the model with just two of the four remaining parameters q_1, q_2, ε_N and S_N, and without violating the boundary conditions. In the context of the present work, it was decided to set the parameters q_1 and q_2 according to the recommendations of Tvergaard and Needleman [17], thus, ensuring better comparability of the model parameters for other studies. Nevertheless, it should be noted that the course of f^* was highly sensitive to influence from q_1 and q_2, which would be advantageous in a representation of the exact course. In the present study, only the three boundary conditions Equations (7)–(9) had to be fulfilled. The set of temperature-dependent GTN model parameters developed for the investigated material is listed in the following Table 2:

Table 2. Compilation of the determined parameters for the GTN model depending on the temperature for AZ31 thin sheet.

ϑ	q_1	q_2	q_3	f_0	f_N	S_N	ε_N	f_c	f_f
RT	1.5	1	2.25	0.0004	0.011	0.0246	0.246	0.0018	0.0047
150 °C						0.0594	0.475	0.0078	0.051
250 °C						0.1188	0.95	0.0089	0.075
350 °C						0.1544	1.235	0.0123	0.104

The determined parameters also reflect the real void nucleation process very well. In Figure 11, it can be seen that f_{nuc} has not yet reached its maximum before it fractures at $f^* = f_f$. This comes remarkably close to the experimental results, because both the analysis of the fracture surfaces and the analysis of the proportion of voids in direct contact with second-phase particles indicated that not all particles contributed to void nucleation. Furthermore, the parameters for the GTN model showed a linear dependence on temperature, which can be represented with good approximation by Equations (10)–(13).

$$S_N = 0.0004 \cdot \vartheta + 0.0107 \quad \left(R^2 = 0.977\right) \tag{10}$$

$$\varepsilon_N = 0.0031 \cdot \vartheta + 0.1249 \quad \left(R^2 = 0.967\right) \tag{11}$$

$$f_c = 0.00003 \cdot \vartheta + 0.0019 \quad \left(R^2 = 0.953\right) \tag{12}$$

$$f_f = 0.0003 \cdot \vartheta + 0.0015 \quad \left(R^2 = 0.993\right) \tag{13}$$

Thus, and analogous to the hardening and softening behavior and anisotropic hardening, a continuously differentiable parameter from RT to 350 °C is available for damage modeling in FEM simulation.

4. Summary and Conclusions

With the use of in-situ SEM tensile tests, twin-roll cast, hot rolled, and annealed magnesium AZ31 sheet was studied with respect to active ductile damage, as well as deformation mechanisms during temperatures ranging from room temperature to 350 °C. Three specimen types were examined for simple tension, high triaxiality, and simple shear stress states. Furthermore, post-mortem analyses of specimen cross-sections were conducted. Here, ductile damage through void nucleation, coalescence, and growth was linked to digital image correlation strain measurements to determine material parameters. Based on the damage model by Gurson–Tvergaard–Needleman (the GTN model), these parameters were established to be used for finite element simulations with AZ31 magnesium (see Table 2). An explicit formulation of the void growth rate has been chosen to prove the functionality of the parameters. From the presented study, the authors conclude:

- SEM in-situ images have shown that the specimen surface shows no characteristic signs of ductile damage, but rather allows statements to be made about the deformation mechanisms;
- Near-surface phenomena of slip traces, grain boundary sliding, and dynamic recrystallization could be observed as functions of temperature and stress state;
- For the first time in literature, an experimental method is proposed and validated to determine all GTN model parameters for sheet metal;
- The post-mortem analyses confirm that pre-existing hydrogen voids, particle-induced voids, as well as voids from grain boundary failure or twin-boundary failure were responsible for ductile damage accumulation. Nevertheless, the post-mortem analyses do not provide sufficient insight into the exact statistics of void origin;
- The established GTN model parameters show a linear positive dependence on temperature. Furthermore, this parameterization represents a milestone in the ductile damage modeling of Mg thin sheet, because the large temperature ranges that can occur in sheet metal forming can be simulated;
- The authors conclude that the anisotropic deformation properties can still be modeled as soon as they are considered in the yield stress of the GTN model.

Author Contributions: Conceptualization, T.H. and S.O.; methodology, T.H. and S.O.; software, T.H.; validation, M.U., S.O. and U.P.; formal analysis, T.H.; investigation, T.H.; resources, S.O., U.P. and R.K.; data curation, T.H.; writing—original draft preparation, T.H. and S.O.; writing—review and editing, M.U., U.P. and R.K.; visualization, T.H.; supervision, U.P. and R.K.; project administration, T.H.; funding acquisition, S.O. and M.U. All authors have read and agreed to the published version of the manuscript.

Funding: The authors thank the German Research Foundation (DFG) for its financial support of the project "Characterisation and modelling of the biaxial material behaviour of twin-roll cast, hot rolled, and annealed AZ31 sheets" (AW6/34-1; UL471/3-1). This research bears the DFG grant number 396576920. Further, S. Osovski acknowledges the financial support provided by the Pazy foundation young researchers award grant number 1176.

Conflicts of Interest: The authors declare no conflict of interest. The funders had no role in the design of the study; in the collection, analyses, or interpretation of data; in the writing of the manuscript, or in the decision to publish the results.

Appendix A

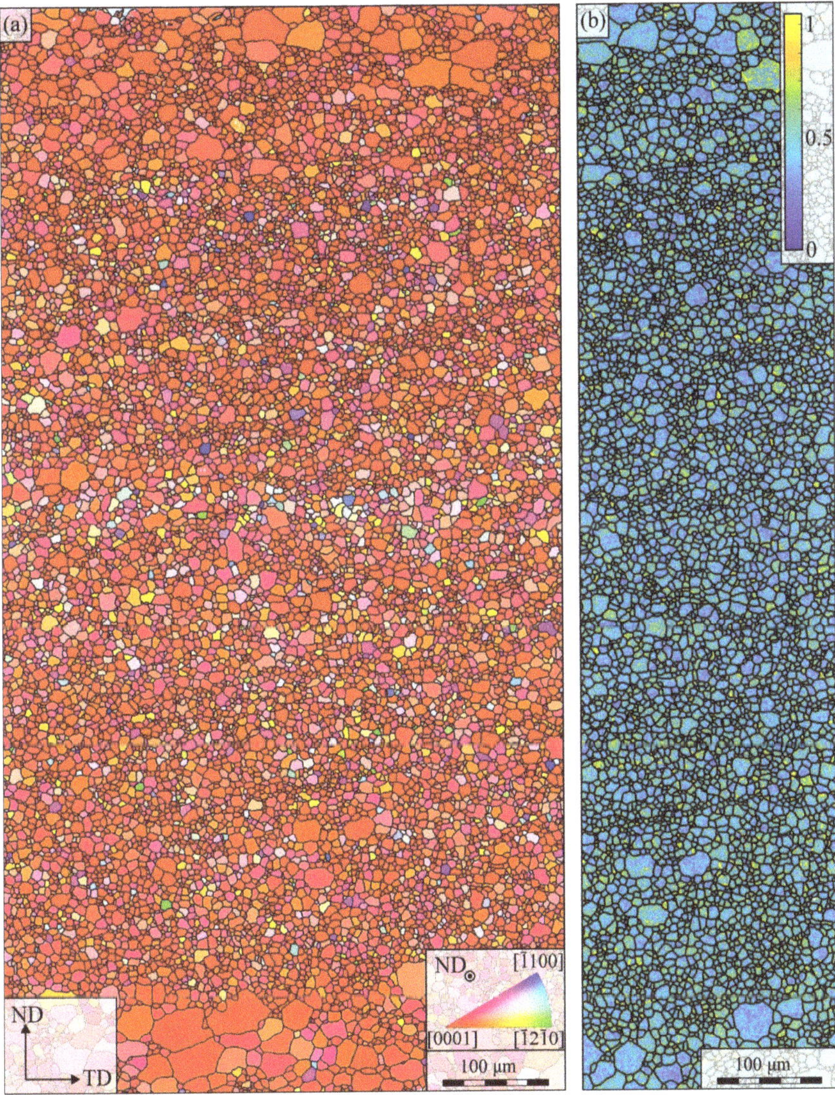

Figure A1. (**a**) Grain orientation map of the specimen cross-section in ND (normal direction)-TD (transverse direction) plane; the color grading indicates the crystal plane normal, which is parallel to ND; (**b**) intragranular misorientation map on the identical area of analysis.

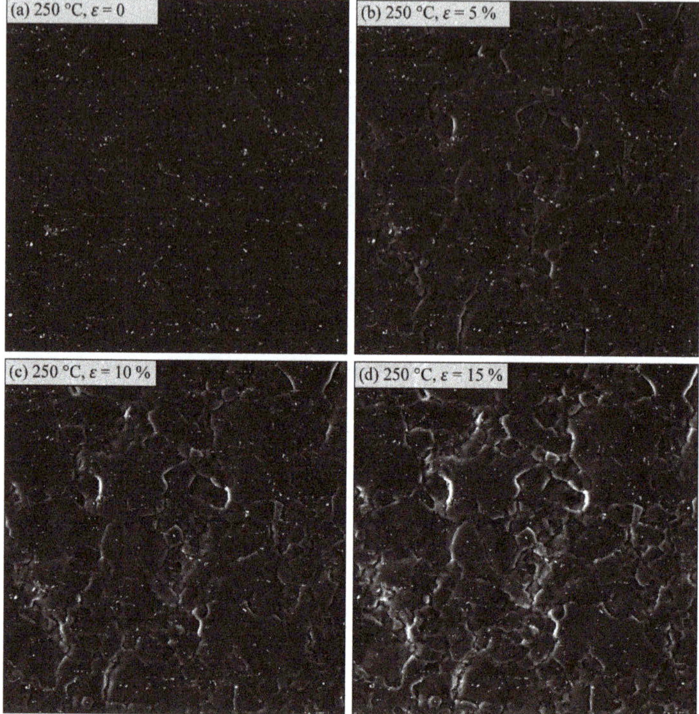

Figure A2. In the undeformed condition, the specimen surface has a dark grey appearance with uniformly distributed second-phase particles in bright contrast; Here, SEM images of (**a**) undeformed sample surface as well as continuously increased strain at (**b**) 5%, (**c**) 10%, and (**d**) 15%, are shown.

References

1. Cao, T. Models for ductile damage and fracture prediction in cold bulk metal forming processes. *Int. J. Mater. Form.* **2017**, *10*, 139–171. [CrossRef]
2. Wu, Z.; Curtin, W.A. The origins of high hardening and low ductility in magnesium. *Nature* **2015**, *526*, 62–67. [CrossRef] [PubMed]
3. John Neil, C.; Agnew, S. Crystal plasticity-based forming limit prediction for non-cubic metals: Application to Mg alloy AZ31B. *Int. J. Plast.* **2009**, *25*, 379–398. [CrossRef]
4. Barnett, M. Influence of deformation conditions and texture on the high temperature flow stress of magnesium AZ31. *J. Light Met.* **2001**, *1*, 167–177. [CrossRef]
5. Barnett, M. A taylor model based description of the proof stress of magnesium AZ31 during hot working. *Metall. Mater. Trans. A* **2003**, *34*, 1799–1806. [CrossRef]
6. Kim, W.; Lee, B. Factors influencing tensile ductility of ultrafine-grained Mg–3Al–1Zn alloy sheet at elevated temperatures. *Mater. Sci. Eng. A* **2010**, *527*, 5984–5989. [CrossRef]
7. Kim, W.; Chung, S.; Chung, C.S.; Kum, D. Superplasticity in thin magnesium alloy sheets and deformation mechanism maps for magnesium alloys at elevated temperatures. *Acta Mater.* **2001**, *49*, 3337–3345. [CrossRef]
8. Nemcko, M.; Wilkinson, D. On the damage and fracture of commercially pure magnesium using x-ray microtomography. *Mater. Sci. Eng. A* **2016**, *676*, 146–155. [CrossRef]
9. Ray, A.; Wilkinson, D. The effect of microstructure on damage and fracture in AZ31B and ZEK100 magnesium alloys. *Mater. Sci. Eng. A* **2016**, *658*, 33–41. [CrossRef]
10. Kondori, B.; Benzerga, A. Modeling damage accumulation to fracture in a magnesium-rare earth alloy. *Acta Mater.* **2017**, *124*, 225–236. [CrossRef]

11. Lhuissier, P.; Scheel, M.; Salvo, L.; Di Michiel, M.; Blandin, J. Continuous characterization by X-ray microtomography of damage during high-temperature deformation of magnesium alloy. *Scr. Mater.* **2013**, *69*, 85–88. [CrossRef]
12. Lhuissier, P.; Scheel, M.; Salvo, L.; Boller, E.; Di Michiel, M.; Blandin, J. 4D Damage Characterization during Superplastic Deformation of Magnesium Alloys. *MSF* **2012**, *735*, 61–66. [CrossRef]
13. Rashed, H.M.M.A.; Robson, J.; Bate, P.; Davis, B. Application of X-ray microtomography to analysis of cavitation in AZ61 magnesium alloy during hot deformation. *Mater. Sci. Eng. A* **2011**, *528*, 2610–2619. [CrossRef]
14. Toda, H.; Shamsudin, Z.; Shimizu, K.; Uesugi, K.; Takeuchi, A.; Suzuki, Y.; Nakazawa, M.; Aoki, Y.; Kobayashi, M. Cavitation during high-temperature deformation in Al–Mg alloys. *Acta Mater.* **2013**, *61*, 2403–2413. [CrossRef]
15. Kondori, B.; Benzerga, A. Effect of Stress Triaxiality on the Flow and Fracture of Mg Alloy AZ31. *Metall. Mater. Trans. A* **2014**, *45*, 3292–3307. [CrossRef]
16. Gurson, A. Continuum Theory of Ductile Rupture by Void Nucleation and Growth. *J. Eng. Mater. Technol.* **1977**, *99*, 2–15. [CrossRef]
17. Tvergaard, V.; Needleman, A. Analysis of the cup-cone fracture in a round tensile bar. *Acta Metall.* **1984**, *32*, 157–169. [CrossRef]
18. Tvergaard, V. Influence of voids on shear band instabilities under plane strain conditions. *Int. J. Fract.* **1981**, *17*, 389–407. [CrossRef]
19. Chu, C.; Needleman, A. Void Nucleation Effects in Biaxially Stretched Sheets. *J. Eng. Mater. Technol.* **1980**, *102*, 249. [CrossRef]
20. Wang, R.; Chen, Z.; Li, Y.; Dong, C. Failure analysis of AZ31 magnesium alloy sheets based on the extended GTN damage model. *Int. J. Miner. Metall. Mater.* **2013**, *20*, 1198–1207. [CrossRef]
21. Zhao, P.; Chen, Z.; Dong, C. Damage and Failure Analysis of AZ31 Alloy Sheet in Warm Stamping Processes. *J. Mater. Eng. Perform.* **2016**, *25*, 2702–2710. [CrossRef]
22. Costa Mattos, H.; Minak, G.; Di Gioacchino, F.; Soldà, A. Modeling the superplastic behavior of Mg alloy sheets under tension using a continuum damage theory. *Mater. Des.* **2009**, *30*, 1674–1679. [CrossRef]
23. Kawalla, R.; Oswald, M.; Schmidt, C.; Ullmann, M.; Vogt, H.-P.; Cuong, N. Development of a strip-rolling technology for Mg alloys based on the twin-roll-casting process. *Miner. Met. Mater. Soc.* **2008**, *177*, 182.
24. Kawalla, R.; Ullmann, M.; Oswald, M.; Schmidt, C. Properties of strips and sheets of magnesium alloy produced by casting-rolling technology. In Proceedings of the 7th International Conference on Magnesium Alloys and Their Applications, Dresden, Germany, 2 January 2007.
25. Henseler, T.; Ullmann, M.; Kawalla, R.; Berge, F. Influence of the Sheet Manufacturing Process on the Forming Limit Behaviour of Twin-Roll Cast, Rolled and Heat-Treated AZ31. *Key Eng. Mater.* **2017**, *746*, 154–160. [CrossRef]
26. Ullmann, M.; Berge, F.; Neh, K.; Kawalla, R. Development of a rolling technology for twin-roll cast magnesium strip. *Metalurgija* **2015**, *4*, 711–714.
27. Graf, M.; Henseler, T.; Ullmann, M.; Kawalla, R.; Prahl, U.; Awiszus, B. Study on determination of flow behaviour of 6060-aluminium and AZ31-magnesium thin sheet by means of stacked compression test. *IOP Conf. Ser. Mater. Sci. Eng.* **2019**, *480*, 12023. [CrossRef]
28. Kawalla, R.; Ullmann, M.; Henseler, T.; Prahl, U. Magnesium Twin-Roll Casting Technology for Flat and Long Products-State of the Art and Future. *Mater. Sci. Forum* **2018**, *941*, 1431–1436. [CrossRef]
29. Bachmann, F.; Hielscher, R.; Schaeben, H. Texture Analysis with MTEX–Free and Open Source Software Toolbox. *SSP* **2010**, *160*, 63–68. [CrossRef]
30. Chiu, S.N.; Stoyan, D.; Kendall, W.S.; Mecke, J. *Stochastic Geometry and Its Applications*; John Wiley & Sons: Hoboken, NJ, USA, 2013.
31. Blaber, J.; Adair, B.; Antoniou, A. Ncorr: Open-Source 2D Digital Image Correlation Matlab Software. *Exp. Mech.* **2015**, *55*, 1105–1122. [CrossRef]
32. Ullmann, M.; Kittner, K.; Henseler, T.; Stöcker, A.; Prahl, U.; Kawalla, R. Development of new alloy systems and innovative processing technologies for the production of magnesium flat products with excellent property profile. *Procedia Manuf.* **2019**, *27*, 203–208. [CrossRef]
33. Masoumi, M.; Zarandi, F.; Pekguleryuz, M. Microstructure and texture studies on twin-roll cast AZ31 (Mg–3wt.%Al–1wt.%Zn) alloy and the effect of thermomechanical processing. *Mater. Sci. Eng. A* **2011**, *528*, 1268–1279. [CrossRef]

34. Liang, P.; Tarfa, T.; Robinson, J.; Wagner, S.; Ochin, P.; Harmelin, M.; Seifert, H.; Lukas, H. Experimental investigation and thermodynamic calculation of the Al–Mg–Zn system. *Thermochim. Acta* **1998**, *314*, 87–110. [CrossRef]
35. Koike, J.; Ohyama, R.; Kobayashi, T.; Suzuki, M.; Maruyama, K. Grain-Boundary Sliding in AZ31 Magnesium Alloys at Room Temperature to 523 K. *Mater. Trans.* **2003**, *44*, 445–451. [CrossRef]
36. Xia, D.; Huang, G.; Liu, S.; Tang, A.; Gavras, S.; Huang, Y.; Hort, N.; Jiang, B.; Pan, F. Microscopic deformation compatibility during biaxial tension in AZ31 Mg alloy rolled sheet at room temperature. *Mater. Sci. Eng. A* **2019**, *756*, 1–10. [CrossRef]
37. Üçel, İ.; Kapan, E.; Türkoğlu, O.; Aydıner, C. In situ investigation of strain heterogeneity and microstructural shear bands in rolled Magnesium AZ31. *Int. J. Plast.* **2019**, *118*, 233–251. [CrossRef]
38. Wang, F.; Sandlöbes, S.; Diehl, M.; Sharma, L.; Roters, F.; Raabe, D. In situ observation of collective grain-scale mechanics in Mg and Mg–rare earth alloys. *Acta Mater.* **2014**, *80*, 77–93. [CrossRef]
39. Xi, B.; Fang, G.; Xu, S. In-situ analysis of microscopic plastic and failure behaviors of extruded magnesium alloy. *Mater. Sci. Eng. A* **2019**, *749*, 148–157. [CrossRef]
40. Wang, H.; Boehlert, C.; Wang, Q.; Yin, D.; Ding, W. In-situ analysis of the tensile deformation modes and anisotropy of extruded Mg-10Gd-3Y-0.5Zr (wt.%) at elevated temperatures. *Int. J. Plast.* **2016**, *84*, 255–276. [CrossRef]
41. Cepeda-Jiménez, C.; Molina-Aldareguia, J.; Pérez-Prado, M. EBSD-Assisted Slip Trace Analysis during in Situ SEM Mechanical Testing: Application to Unravel Grain Size Effects on Plasticity of Pure Mg Polycrystals. *JOM* **2016**, *68*, 116–126. [CrossRef]
42. Osovski, S.; Srivastava, A.; Williams, J.; Needleman, A. Grain boundary crack growth in metastable titanium β alloys. *Acta Mater.* **2015**, *82*, 167–178. [CrossRef]
43. Liu, Y.; Zheng, X.; Osovski, S.; Srivastava, A. On the micromechanism of inclusion driven ductile fracture and its implications on fracture toughness. *J. Mech. Phys. Sol.* **2019**, *130*, 21–34. [CrossRef]
44. Wang, D.; Shanthraj, P.; Springer, H.; Raabe, D. Particle-induced damage in Fe–TiB2 high stiffness metal matrix composite steels. *Mater. Des.* **2018**, *160*, 557–571. [CrossRef]
45. Barnett, M. Twinning and the ductility of magnesium alloys. *Mater. Sci. Eng. A* **2007**, *464*, 8–16. [CrossRef]
46. Ulacia, I.; Yi, S.; Hurtado, I. High strain rate formability of AZ31B magnesium alloy sheets. In *Magnesium: Proceedings of the 8th International Conference on Magnesium Alloys and Their Applications*; Kainer, K.U., Ed.; Wiley-VCH: Weinheim, Germany, 2010; pp. 509–515.
47. Bieler, T.; Eisenlohr, P.; Roters, F.; Kumar, D.; Mason, D.; Crimp, M.; Raabe, D. The role of heterogeneous deformation on damage nucleation at grain boundaries in single phase metals. *Int. J. Plast.* **2009**, *25*, 1655–1683. [CrossRef]
48. Uthaisangsuk, V.; Prahl, U.; Bleck, W. Micromechanical modelling of damage behaviour of multiphase steels. *Comput. Mater. Sci.* **2008**, *43*, 27–35. [CrossRef]
49. Benzerga, A.; Leblond, J.-B. Ductile fracture by void growth to coalescence. *Adv. Appl. Mech.* **2010**, 169–305.
50. Nemcko, M.; Li, J.; Wilkinson, D. Effects of void band orientation and crystallographic anisotropy on void growth and coalescence. *J. Mech. Phys. Sol.* **2016**, *95*, 270–283. [CrossRef]
51. Landron, C.; Maire, E.; Adrien, J.; Bouaziz, O. Damage characterization in Dual-Phase steels using X-ray tomography. In *Optical Measurements, Modeling, and Metrology*; Proulx, T., Ed.; Springer: New York, NY, USA, 2011; Volume 5, pp. 11–18.
52. Tvergaard, V. Interaction of very small voids with larger voids. *Int. J. Solids Struct.* **1998**, *35*, 3989–4000. [CrossRef]
53. Prahl, U.; Rehbach, W.; Kuckertz, C.; Weichert, D.; Bleck, W. Ductile Damage Prediction on the Basis of Microstructural Observations. *KEM* **2003**, *251–252*, 351–356. [CrossRef]
54. Prahl, U. Schädigung und Versagen Mikrolegierter Feinkornstähle in Experiment und Simulation. Ph.D. Thesis, RWTH Aachen University, Aachen, Germany, 2002.

Article

Microstructure Evolution and Mechanical Properties of AZ31 Magnesium Alloy Sheets Prepared by Low-Speed Extrusion with Different Temperature

Wenyan Zhang [1], Hua Zhang [1,*], Lifei Wang [2], Jianfeng Fan [2], Xia Li [1], Lilong Zhu [1], Shuying Chen [1], Hans Jørgen Roven [3] and Shangzhou Zhang [1]

1 Institute for Advanced Studies in Precision Materials, Yantai University, Yantai 264005, China; jzcl@ytu.edu.cn (W.Z.); xym_007@126.com (X.L.); lilong.zhu@ytu.edu.cn (L.Z.); sychen2014@gmail.com (S.C.); szzhangyt@163.com (S.Z.)

2 Department of Materials Science and Engineering, Taiyuan University of Technology, Taiyuan 030024, China; wanglifei@tyut.edu.cn (L.W.); fanjianfeng@tyut.edu.cn (J.F.)

3 Department of Materials Science and Engineering, Norwegian University of Science and Technology, 7491 Trondheim, Norway; hans.roven@material.ntnu.no

* Correspondence: zhanghua2009@126.com; Tel.: +86-535-6902856

Received: 28 June 2020; Accepted: 21 July 2020; Published: 26 July 2020

Abstract: AZ31 magnesium alloy sheets were prepared by low-speed extrusion at different temperatures, i.e., 350 °C, 400 °C, and 450 °C. The microstructure evolution and mechanical properties of extruded AZ31 magnesium alloy sheets were studied. Results indicate that the low-speed extrusion obviously improved the microstructure of magnesium alloys. As the extrusion temperature decreased, the grain size for the produced AZ31 magnesium alloy sheets decreased, and the (0001) basal texture intensity of the extruded sheets increased. The yield strength and tensile strength of the extruded sheets greatly increased as the extrusion temperature decreased. The AZ31 magnesium alloy sheet prepared by low-speed extrusion at 350 °C exhibited the finest grain size and the best mechanical properties. The average grain size, yield strength, tensile strength, and elongation of the extruded sheet prepared by low-speed extrusion at 350 °C were ~2.7 μm, ~226 MPa, ~353 MPa, and ~16.7%, respectively. These properties indicate the excellent mechanical properties of the extruded sheets prepared by low-speed extrusion. The grain refinement effect and mechanical properties of the extruded sheets produced in this work were obviously superior to those of magnesium alloys prepared using traditional extrusion or rolling methods reported in other related studies.

Keywords: magnesium alloy; low-speed extrusion; microstructure evolution; mechanical properties

1. Introduction

At present, magnesium alloy products have been used in almost every aspect of human life. For example, magnesium alloys can be used to produce light-weight automobile components or to satisfy the noise-absorbing, shock-absorbing, and radiation-resistance requirements in the aerospace industry [1,2]. Magnesium alloys can also be used to build lighter weapons and equipment for military use, which will enhance maneuverable tactical performance and performances in remote and precision strikes. Further, magnesium alloys have also been used widely in electrical and information industries due to their advantages, e.g., the high specific strength, good thermal and electrical conductivity, environmental friendliness, and excellent electromagnetic shielding performance, and because they meet the requirements for light, thin, and small-footprint 3C products [2]. Therefore, accelerated research and development of magnesium alloy materials will have a profound impact on the promotion of societal development and human life improvement.

Currently, the majority of magnesium alloy products are castings, especially die-castings [2,3]. However, the microstructures of such products are not ideal and the mechanical properties are poor, which has greatly hindered the wider applications of magnesium alloys. The deformed magnesium alloys prepared by plastic processing typically exhibit higher strength and toughness, and can thus meet the requirements of a good deal of structural components [4,5]. Forward extrusion is a widely used plastic processing method in the production of wrought magnesium alloys. During the extrusion, the material will be subjected to high hydrostatic pressure, and a large amount of deformation will occur, which is beneficial for eliminating microstructure defects in the ingots, thus effectively improving the comprehensive properties of the material [6]. In comparison to other plastic processing methods, e.g., forging and rolling, the extrusion process has the following advantages. First, the extrusion process is highly flexible and easy to perform. Second, the precision in terms of the size of the extruded products and the surface quality is high. Therefore, investigations of extruded magnesium alloys have been the subject of research. Han et al. [7] prepared AZ31 magnesium alloy sheets with a fine grain size of 1.4 μm by accumulated extrusion at 200 °C with an extrusion speed of 6 mm/s and an extrusion ratio of 12.8:1. Yang et al. [8] studied microstructure evolution of AZ31 magnesium alloys extruded by a high extrusion ratio of 94:1 at 430 °C, and an inhomogeneous microstructure with fine grains (~8 μm) and coarse grains (~25 μm) was obtained. Jiang et al. [9] studied microstructure evolution of AZ31 magnesium alloy extruded at 300 °C with an extrusion speed of 10 mm/s and revealed the generation of basal texture variations. However, most studies have mainly focused on understanding the microstructure and mechanical properties of magnesium alloys produced by extrusion with a relatively high extrusion speed [7–9]. Based on previous studies, in the present work, the effects of low-speed extrusion on the microstructure evolution and mechanical properties of AZ31 magnesium alloys under different extrusion temperatures were investigated, with the aim of producing AZ31 magnesium alloy sheets with fine microstructure and excellent mechanical properties.

2. Experiment Procedures

The extrusion temperature of magnesium and its alloys is usually 300–450 °C. To understand the effect of low-speed extrusion on the microstructure and mechanical properties of AZ31 magnesium alloys, the microstructure evolution and mechanical properties of the sheets prepared by low-speed extrusion at 350 °C, 400 °C, and 450 °C were investigated. Commercially available AZ31 magnesium alloy rods having a diameter of 40 mm were applied as original samples. The DK7720 electrical-discharge wire-cutting machine (Taizhou Huadong CNC Machine Tool Co. LTD, Taizhou, China) was used to process the original samples into cylindrical samples having a diameter of 40 mm and a height of 30 mm. The cylindrical samples were then annealed in the KSL-1200X heating box furnace (Hefei Kejing Material Technology Co. LTD, Hefei, China) at 400 °C for 2 h and subsequently extruded at 350 °C, 400 °C, and 500 °C on a hydraulic press (Jinan Zhonglu Chang Testing Machine Manufacturing Co. LTD, Jinan, China), respectively. Prior to extrusion, the extrusion mold placed in a hollow furnace was heated to the extrusion temperature and maintained for 2 h. The wire-cut cylindrical samples were then placed in the mold and maintained for 1.5 h under the extrusion temperature. Low-speed extrusion was then performed, and the extruded samples were cooled instantly in water after extrusion. An extrusion rate of 1 mm/s and an extrusion ratio of 16:1 were applied in this work. The extruded magnesium alloy sheets were 30 mm in width and 2 mm in thickness. Optical microscopy (Keenz Corporation, Osaka, Japan) and electron back scattering diffraction (EBSD) (Oxford Instruments, Oxford, England) installed on a FIB/ SEM double-beam electron microscope (TESCAN, Brno, Czech Republic) were applied to study the microstructure evolution of the extruded magnesium alloy sheets. The electrical-discharge wire-cutting machine was used to cut the extruded sheets along the extrusion direction. The bone-shaped tensile samples had gauge dimensions of 10 mm (length) × 3 mm (width) × 2 mm (thickness). The tensile testing was conducted using an AGX-XD-30 κN electronic universal testing machine (Shimazu Instrument (Suzhou) Co. LTD, Suzhou, China) at ambient temperature, where the tensile strain rate was 1×10^{-3} s^{-1}. Three tensile tests were performed for each extruded condition.

3. Results and Discussion

3.1. Microstructure of Original AZ31 Magnesium Alloy Rod

As indicated by Figure 1, the microstructure of the original magnesium alloy rod was very inhomogeneous. Coarse equiaxed grains, fine equiaxed grains, and elongated grains all existed in the original magnesium alloy rod. This is because the deformation degree of the original magnesium alloy bar cannot reach the conditions of complete recrystallization during production. The deformation amount in some areas is below the critical deformation degree, where dynamic recrystallization cannot occur, which results in the existence of coarse equiaxed and elongated grains, i.e., incomplete dynamic recrystallization [10,11]. Figure 2 presents the EBSD map with the (0002), $(10\bar{1}0)$, and $(11\bar{2}0)$ pole figures of the original rod. Coarse equiaxed grains, fine equiaxial grains, and elongated grains were all observed within the original magnesium alloy rod, indicating an inhomogeneous microstructure, as in the results of metallographic microstructure analyses. As indicated by the (0002) pole figure in Figure 2, a typical annular extruded fiber texture was observed in the original magnesium alloy bar, where the (0002) basal planes in the majority of grains were parallel to the extrusion direction, and the max intensity of basal texture was 15.08. It can be seen from Figure 2 that the preferred orientations of the $(10\bar{1}0)$ and $(11\bar{2}0)$ prismatic planes in the original AZ31 magnesium alloy bar are not obvious. The $(10\bar{1}0)$ pole figure indicates that the $(10\bar{1}0)$ prismatic plane is parallel to the extrusion direction.

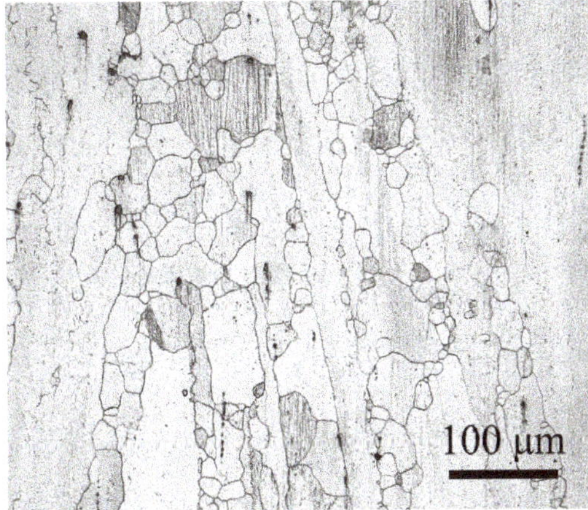

Figure 1. Metallographic microstructure of original AZ31 magnesium alloy rod.

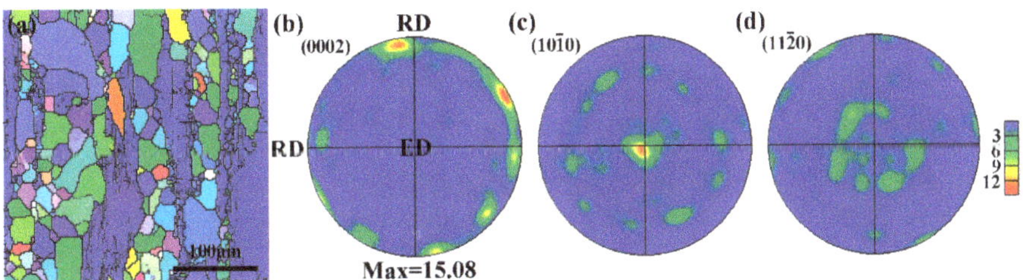

Figure 2. (**a**) Electron back scattering diffraction (EBSD) map and (**b**) (0002), (**c**) $(10\bar{1}0)$, (**d**) $(11\bar{2}0)$ pole figures of original AZ31 magnesium alloy rod.

3.2. Microstructure of AZ31 Magnesium Alloy Sheets Prepared by Low-Speed Extrusion

Figure 3 presents the positions for microstructure observation of the extruded sheet prepared by low-speed extrusion. Figure 4 presents the metallographic microstructure of the extruded sheets prepared by low-speed extrusion at different temperatures, where Figure 4a–c indicates the metallographic microstructure at position 1 suggested by Figure 3, and Figure 4d–f shows the metallographic microstructure at position 2. As indicated by Figure 4a–c, at position 1 of the extruded sheets, dynamic recrystallization occurred at the grain boundary and a ring-shaped fine grain microstructure formed. This phenomenon became more obvious for the sheets extruded at 350 °C and 400 °C. The deformation amount of the magnesium alloy was minor at position 1, and the stored deformation energy was also insufficient such that partial original coarse grains and elongated deformed microstructure were still preserved. Because of the high degree of plastic deformation and dramatic dynamic recrystallization, the microstructure of the extruded sheets prepared by low-speed extrusion under different temperatures became more homogeneous at position 2, and the average grain size was reduced. Fine equiaxed grains formed via dynamic recrystallization in the extruded sheets prepared by low-speed extrusion. During the dynamic recrystallization process, the average size (***d***) of recrystallized grains has the following relationship with the Zener–Hollomon parameter (Z): $\ln d = A - B\ln Z$. Further, Z, which represents the temperature-modified strain rate, is $Z = \varepsilon \exp(Q/RT)$ [12]. Here, A and B are constants, ε denotes the strain rate during extrusion, Q denotes the deformation activation energy, T denotes the extrusion temperature, and R is the gas constant. It is concluded from the above formula that the size for dynamic recrystallized grains is inversely proportional to $\ln Z$ and that Z is also inversely proportional to temperature T. Therefore, the average size of grains formed by dynamic recrystallization increased as the extrusion temperature rose. As indicated by Figure 4d–f, with increasing extrusion temperature, the grain size of the extruded sheets prepared by low-speed extrusion at position 2 increased. The average grain sizes of the AZ31 magnesium alloy sheets extruded at 350 °C, 400 °C, and 450 °C were 2.7 ± 0.2 μm, 4.1 ± 0.3 μm, and 6.0 ± 0.3 μm, respectively. Thus, the influence of low-speed extrusion on the grain refinement of AZ31 Mg alloy sheets was dramatic. Furthermore, the grain size of ~2.7 μm for the extruded sheets produced by low-speed extrusion at 350 °C was substantially lower than the average grain size of extruded AZ31 magnesium alloys reported in the literature [13–15]. Li et al. [13] suggested that the average grain size of AZ31 alloy sheets prepared by continuous variable cross-section direct extrusion at 350 °C, an extrusion speed of 1 mm/s, and an extrusion ratio of 44.4 was 6.18 μm. Xu et al. [14] prepared AZ31 magnesium alloys by forward extrusion having an extrusion ratio of 33:1 and an extrusion temperature of 400 °C, and the average grain size was 9.3 μm. Pan et al. [15] prepared AZ31 magnesium alloys by conventional direct extrusion with an extrusion ratio of 31.52 at 400 °C, and the microstructure was very inhomogenous, which consisted of 6–8 μm fine grains and 13–16 μm irregular elongated grains.

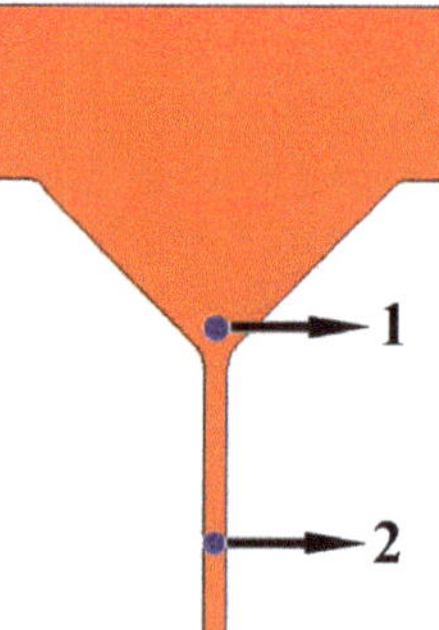

Figure 3. Positions for microstructure observation of the extruded sheet prepared by low-speed extrusion.

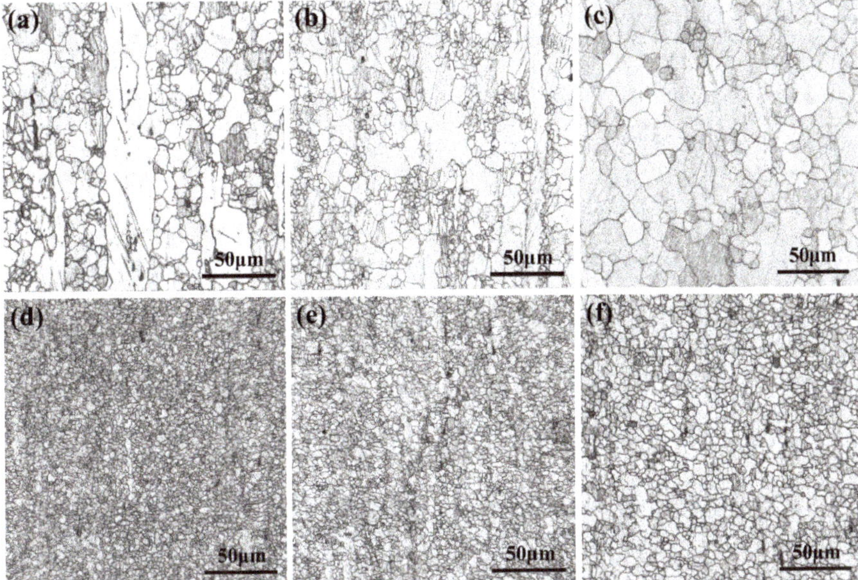

Figure 4. Metallographic microstructure of the extruded sheets prepared by low-speed extrusion at different temperatures: (**a**,**d**) 350 °C; (**b**,**e**) 400 °C; (**c**,**f**) 450 °C.

Figure 5 presents the EBSD maps and (0002), $(10\bar{1}0)$, and $(10\bar{2}0)$ pole figures at position 2 of the extruded sheets produced by low-speed extrusion at different temperatures. The EBSD maps in Figure 5 also show that the extruded sheets prepared by low-speed extrusion had a uniform microstructure, and the grain size increased as the extrusion speed rose; the results are similar to those in Figure 4. As indicated by the (0002) pole figure in Figure 5, all the extruded sheets produced by low-speed extrusion exhibited a typical (0002) basal texture, and the (0002) basal planes in the majority of grains were parallel to the extrusion direction. Further, with increasing extrusion temperature, the max basal texture intensity of the extruded sheets decreased gradually. The max basal texture intensity of the extruded sheets produced by extrusion at 350 °C, 400 °C, and 450 °C was 18.03, 16.01, and 15.82, respectively. This was because as the extrusion temperature rose, the critical resolved shear stress (CRSS) of the non-basal slips, such as prismatic slip and pyramidal slip, of the magnesium alloys decreased. For instance, for magnesium alloys, the CRSS of the prismatic slip was close to that of the basal slip, when the deformation temperature was above 300 °C [16,17]. Hence, the initiation of more non-basal slips, e.g., prismatic slip and pyramidal slip, will lead to the increase of grains having a non-basal orientation, which results in the relative decrease of basal texture strength [18,19]. In addition, as indicated by the $(10\bar{1}0)$ and $(10\bar{2}0)$ pole figures in Figure 5, there was no preferred orientation for the $(10\bar{1}0)$ and $(10\bar{2}0)$ prismatic planes in the extruded sheets prepared by low-speed extrusion.

3.3. Mechanical Properties of AZ31 Magnesium Alloy Sheets Prepared by Low-Speed Extrusion

Figure 6 indicates the true tensile stress–strain curves of AZ31 magnesium alloy sheets fabricated by low-speed extrusion at different temperatures. The true stress–strain curves were determined along the extrusion direction under room temperature. Table 1 presents their mechanical properties, i.e., the yield strength, tensile strength, and elongation. As indicated in Figure 6, as the extrusion temperature decreased, the tensile stress–strain curve of the extruded sheets increased gradually, and the yield strength and tensile strength also increased gradually, but the elongation changed slightly. As indicated by Table 1, the extruded sheet obtained by low-speed extrusion at 350 °C exhibited the highest overall mechanical properties. Its yield strength, tensile strength, and elongation were ~226 MPa, ~353 MPa, and ~16.7%, respectively, which are much larger than those of the AZ31 alloy sheets prepared using the traditional extrusion or rolling processes reported in the literature [20–22]. For example, Yang et al. [20] prepared AZ31 alloy sheets by the forward extrusion process at 430 °C having an extrusion ratio of 10:1 and an extrusion rate of 20 mm/s; the extruded magnesium alloy sheet had a yield strength, tensile strength, and elongation of ~156 MPa, ~327 MPa, and ~19.7%, respectively. Li et al. [21] prepared AZ31 alloy sheets at the extrusion temperature of 380 °C, and the yield strength, tensile strength, and elongation of the sheets were 107 MPa, 230 MPa, and 19.3%, respectively. Zhou et al. [22] prepared AZ31 alloy sheets using the traditional rolling process with different rolling speed, and the prepared sheets rolled at a rolling speed of 0.1 mm/s exhibited a yield strength, tensile strength, and elongation of 174 MPa, 279 MPa, and 18.9%, respectively. In addition, as the extrusion temperature rose, the mechanical properties of the extruded sheets prepared by low-speed extrusion deteriorated. The yield strength, tensile strength, and elongation of the sheet extruded at 400 °C were ~200 MPa, ~332 MPa, and ~16.4%, respectively, whereas at the extrusion temperature of 450 °C, they were ~177 MPa, ~320 MPa, and ~17.3%, respectively.

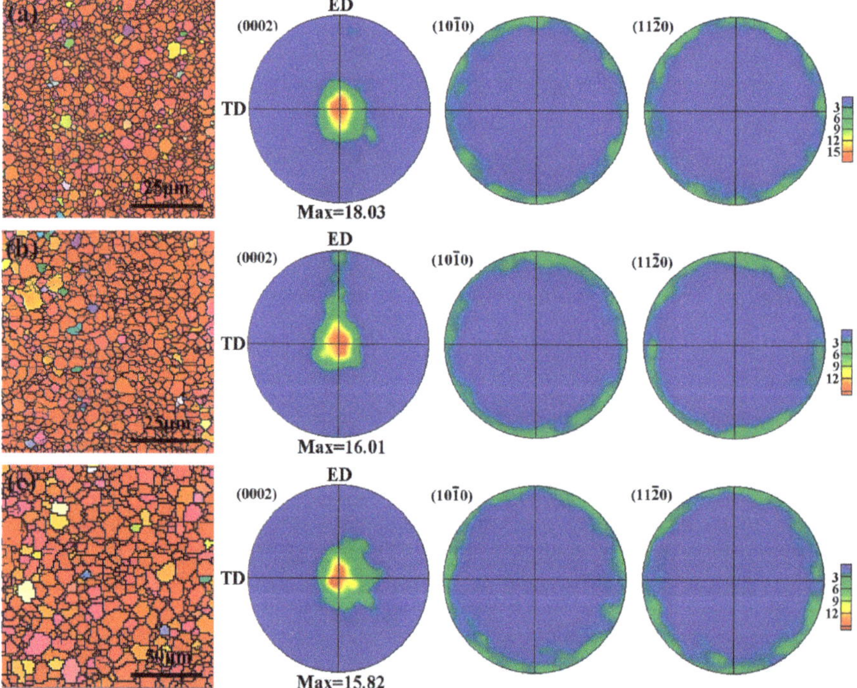

Figure 5. EBSD maps and (0002), (1$\bar{0}$10), (1$\bar{0}$ 20) pole figures of the extruded sheets prepared by low-speed extrusion at different temperatures: (**a**) 350 °C; (**b**) 400 °C; (**c**) 450 °C.

Table 1. Mechanical properties of the extruded sheets fabricated by low-speed extrusion.

Extrusion Temperature	Yield Strength (MPa)	Tensile Strength (MPa)	Elongation (%)
350 °C	226 ± 3	353 ± 3	16.7 ± 0.2
400 °C	200 ± 2	332 ± 3	16.4 ± 0.4
450 °C	177 ± 2	320 ± 4	17.3 ± 0.3

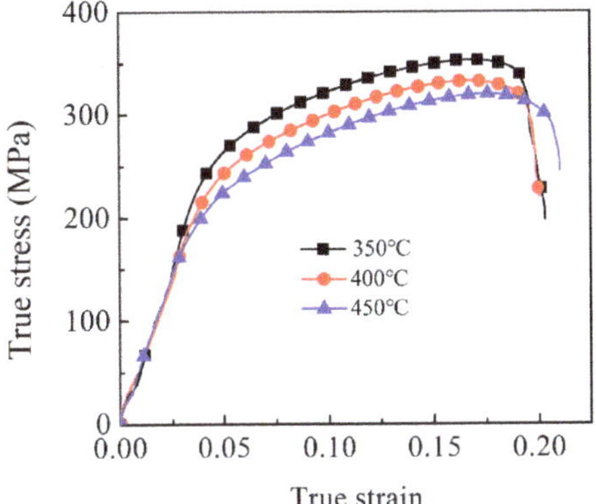

Figure 6. True tensile stress–strain curves of the extruded sheets fabricated by low-speed extrusion.

As indicated by the above results of microstructure analyses, a higher extrusion temperature led to a larger grain size of the extruded sheets prepared by low-speed extrusion, while the grain size had a profound effect on the mechanical properties of metallic materials. Because the grain boundary will hinder the movement of dislocations, the stress concentration generated by the piling up of dislocations in large grains is more likely to induce the occurrence of plastic deformation in adjacent grains compared to that in fine grains. The stress concentration in fine grains is less severe, so a large external stress is required to activate the deformation in adjacent grains [23,24]. Hence, the extruded sheets prepared at 350 °C, having a finer grain size, exhibited a higher strength. In addition, the basal texture strength also has an impact on the yield strength of magnesium alloys. Weak basal texture strength favors the initiation of basal slips during plastic deformation [25], resulting in a low yield strength. This is one of the reasons why the yield strength of the produced sheets decreased with increasing extrusion temperature.

4. Conclusions

This work investigated the effect of the extrusion temperature on the microstructure evolution and mechanical properties of AZ31 magnesium alloy sheets prepared by low-speed extrusion. Results indicated that low-speed extrusion led to the effective refinement of microstructure and enhancement of the mechanical properties of the AZ31 magnesium alloys. When the extrusion temperatures were 350 °C, 400 °C, and 450 °C, the average grain sizes of the extruded sheets were ~2.7 μm, ~4.1 μm, and ~6.0 μm, respectively. A strong (0002) basal texture was observed within the prepared sheets, and the max basal texture intensity was reduced as the extrusion temperature rose. Further, the mechanical properties of the sheets prepared by low-speed extrusion deteriorated as the extrusion temperature rose. When the extrusion temperature was 350 °C, the highest overall mechanical properties were observed, where the yield strength, tensile strength, and elongation were ~226 MPa, ~353 MPa, and ~16.7%, respectively.

Author Contributions: Conceptualization, W.Z., H.Z.; investigation, W.Z., L.W.; formal analysis, J.F., X.L.; writing—original draft, W.Z., H.Z.; validation, L.Z., S.C.; writing—review and editing H.J.R., S.Z. All authors have read and agreed to the published version of the manuscript.

Funding: The authors acknowledge the financial support by National Natural Science Foundation of China (U1810122, U1710118, 51504162 and 51704209), Key Research and Development Program of Shanxi Province (201903D421076 and 201803D421086), Talent Training Program for Shandong Province Higher Educational Youth Innovative Teams (2019), Innovative Talents of Higher Education Institutions of Shanxi (2018), Top Discipline in Materials Science of Shandong Province (2019), and Major Scientific and Technological Innovation Project in Shandong Province (2019JZZY010325).

Conflicts of Interest: The authors declare no conflict of interest.

References

1. Jin, Z.Z.; Cheng, X.M.; Zha, M.; Rong, J.; Zhang, H.; Wang, J.G.; Wang, C.; Li, Z.G.; Wang, H.Y. Effects of $Mg_{17}Al_{12}$ second phase particles on twinning-induced recrystallization behavior in Mg-Al-Zn alloys during gradient hot rolling. *J. Mater. Sci. Technol.* **2019**, *35*, 2017–2026. [CrossRef]
2. Mordike, B.L.; Ebert, T. Magnesium: Properties—Applications—Potential. *Mater. Sci. Eng. A* **2001**, *302*, 37–45. [CrossRef]
3. Zhang, H.; Huang, G.S.; Fan, J.F.; Roven, H.J.; Xu, B.S.; Dong, H.B. Deep drawability and drawing behaviour of AZ31 alloy sheets with different initial texture. *J. Alloys Compd.* **2014**, *615*, 302–310. [CrossRef]
4. Zha, M.; Zhang, H.M.; Yu, Z.Y.; Zhang, X.H.; Meng, X.T.; Wang, H.Y.; Jiang, Q.C. Bimodal microstructure—A feasible strategy for high-strength and ductile metallic materials. *J. Mater. Sci. Technol.* **2018**, *34*, 257–264. [CrossRef]
5. Zeng, Z.R.; Stanford, N.; Davies, C.H.J.; Nie, J.F.; Birbilis, N. Magnesium extrusion alloys: A review of developments and prospects. *Int. Mater. Rev.* **2019**, *64*, 27–62. [CrossRef]
6. Aghion, E.; Bronfin, B. Magnesium alloys development towards the 21st century. *Mater. Sci. Forum* **2000**, *350*, 19–30. [CrossRef]
7. Han, T.Z.; Huang, G.S.; Ma, L.F.; Wang, G.G.; Wang, L.F.; Pan, F.S. Evolution of microstructure and mechanical properties of AZ31 Mg alloy sheets processed by accumulated extrusion bonding with different relative orientation. *J. Alloys Compd.* **2019**, *784*, 584–591. [CrossRef]
8. Yang, Q.S.; Jing, B.; Song, B.; Yu, D.L.; Chai, S.S.; Zhang, J.Y.; Pan, F.S. Mechanical behavior and microstructure evolution for extruded AZ31 sheet under side direction strain. *Prog. Nat. Sci. Mater. Int.* **2020**, *30*, 270–277. [CrossRef]
9. Jiang, M.G.; Xu, C.; Yan, H.; Fan, G.H.; Nakata, T.; Lao, C.S.; Chen, R.S.; Kamado, S.; Han, E.H.; Lu, B.H. Unveiling the formation of basal texture variations based on twinning and dynamic recrystallization in AZ31 magnesium alloy during extrusion. *Acta Mater.* **2018**, *157*, 53–71. [CrossRef]
10. Zhang, Y.; Chen, X.Y.; Liu, Y.L. Microstructure and mechanical properties of as-extruded Mg-Sn-Zn-Ca alloy with different extrusion ratios. *Trans. Nonferrous Met. Soc. China* **2018**, *28*, 2190–2198. [CrossRef]
11. Zhang, X.B.; Yuan, G.Y.; Niu, J.L.; Fu, P.H.; Ding, W.J. Microstructure, mechanical properties, biocorrosion behavior, and cytotoxicity of as-extruded Mg-Nd-Zn-Zr alloy with different extrusion ratios. *J. Mech. Behav. Biomed. Mater.* **2012**, *9*, 153–162. [CrossRef] [PubMed]
12. Ammouri, A.H.; Kridli, G.; Ayoub, G.; Hamade, R.F. Relating grain size to the Zener-Hollomon parameter for twin-roll-cast AZ31B alloy refined by friction stir processing. *J. Mater. Process. Technol.* **2015**, *222*, 301–306. [CrossRef]
13. Li, X.B.; Li, F.; Li, X.W. Microstructural characteristics of AZ31 magnesium alloy processed by continuous variable cross-section direct extrusion (CVCDE). Part 1: Texture evolution. *JOM* **2018**, *70*, 2327–2331. [CrossRef]
14. Xu, J.; Song, J.F.; Jiang, B.; He, J.J.; Wang, Q.H.; Liu, B.; Huang, G.S.; Pan, F.S. Effect of effective strain gradient on texture and mechanical properties of Mg-3Al-1Zn alloy sheets produced by asymmetric extrusion. *Mater. Sci. Eng. A* **2017**, *706*, 172–180. [CrossRef]
15. Pan, F.S.; Wang, Q.H.; Jiang, B.; He, J.J.; Chai, Y.F.; Xu, J. An effective approach called the composite extrusion to improve the mechanical properties of AZ31 magnesium alloy sheets. *Mater. Sci. Eng. A* **2016**, *655*, 339–345. [CrossRef]
16. Yin, D.D.; Wang, W.D.; Boehlert, C.J.; Chen, Z.; Li, H.; Mishra, R.K.; Chakkedath, A. In-Situ Study of the Tensile Deformation and Fracture Modes in Peak-Aged Cast Mg-11Y-5Gd-2Zn-0.5Zr. *Metall. Mater. Trans. A* **2016**, *47*, 6438–6452. [CrossRef]
17. Barnett, M.; Keshavarz, Z.; Beer, A.; Atwell, D. Influence of grain size on the compressive deformation of wrought Mg-3Al-1Zn. *Acta Mater.* **2004**, *52*, 5093–5103. [CrossRef]
18. Agnew, S.R.; Duygulu, O. Plastic anisotropy and the role of non-basal slip in magnesium alloy AZ31B. *Int. J. Plast.* **2005**, *21*, 1161–1193. [CrossRef]
19. Wu, S.K.; Chou, T.S.; Wang, J.Y. The deformation textures in an AZ31B magnesium alloy. *Mater. Sci. Forum* **2003**, *419–422*, 527–532. [CrossRef]
20. Yang, Q.S.; Jiang, B.; Yu, Z.J.; Dai, Q.W.; Luo, S.Q. Effect of extrusion strain path on microstructure and properties of AZ31 magnesium alloy sheet. *Acta Metall. Sin. Engl. Lett.* **2015**, *28*, 1257–1263. [CrossRef]

21. Li, R.H.; Pan, F.S.; Jiang, B.; Dong, H.W.; Yang, Q.S. Effect of Li addition on the mechanical behavior and texture of the as-extruded AZ31 magnesium alloy. *Mater. Sci. Eng. A* **2013**, *562*, 33–38. [CrossRef]
22. Zhou, T.; Yang, Z.; Hu, D.; Feng, T.; Yang, M.B.; Zhai, X.B. Effect of the final rolling speeds on the stretch formability of AZ31 alloy sheet rolled at a high temperature. *J. Alloys Compd.* **2015**, *650*, 436–443. [CrossRef]
23. Apps, P.J.; Karimzadeh, H.; King, J.F.; Lorimer, G.W. Precipitation reactions in Magnesium-rare earth alloys containing Yttrium, Gadolinium or Dysprosium. *Scr. Mater.* **2003**, *48*, 1023–1028. [CrossRef]
24. Dudamell, N.V.; Ulaci, L.; Gálvez, F.; Yi, S.; Bohlen, J.; Letzig, D.; Hurtado, L.; Pérez-Prado, M.T. Twinning and grain subdivision during dynamic deformation of a Mg AZ31 sheet alloy at room temperature. *Acta Mater.* **2011**, *59*, 6949–6962. [CrossRef]
25. Huang, X.S.; Suzuki, K.; Saito, N. Textures and stretch formability of Mg-6Al-1Zn magnesium alloy sheets rolled at high temperatures up to 793 K. *Scr. Mater.* **2009**, *60*, 651–654. [CrossRef]

Article

Thermomechanical Processing of AZ31-3Ca Alloy Prepared by Disintegrated Melt Deposition (DMD)

Kamineni Pitcheswara Rao [1,*], Kalidass Suresh [1,†], Yellapregada Venkata Rama Krishna Prasad [1,‡] and Manoj Gupta [2]

1 Department of Biomedical Engineering, City University of Hong Kong, Tat Chee Avenue, Kowloon, Hong Kong, China; ksureshphy@buc.edu.in (K.S.); prasad_yvrk@hotmail.com (Y.V.R.K.P.)
2 Department of Mechanical Engineering, National University of Singapore, 9 Engineering Drive 1, Singapore 117575, Singapore; mpegm@nus.edu.sg
* Correspondence: mekprao@cityu.edu.hk; Tel.: +852-3442-8409
† Current address: Department of Physics, Bharathiar University, Coimbatore 641046, India.
‡ Current address: Independent Researcher, No. 2/B, Vinayaka Nagar, Hebbal, Bengaluru 560024, India.

Received: 25 June 2020; Accepted: 23 July 2020; Published: 27 July 2020

Abstract: Mg-3Zn-1Al (AZ31) alloy is a popular wrought alloy, and its mechanical properties could be further enhanced by the addition of calcium (Ca). The formation of stable secondary phase $(Mg,Al)_2Ca$ enhances the creep resistance at the expense of formability and, therefore, necessitates the establishment of safe working window(s) for producing wrought products. In this study, AZ31-3Ca alloy has been prepared by the disintegrated melt deposition (DMD) processing route, and its hot deformation mechanisms have been evaluated, and compared with similarly processed AZ31, AZ31-1Ca and AZ31-2Ca magnesium alloys. DMD processing has refined the grain size to 2–3 μm. A processing map has been developed for the temperature range 300–450 °C and strain rate range 0.0003–10 s^{-1}. Three working domains are established in which dynamic recrystallization (DRX) readily occurs, although the underlying mechanisms of DRX differ from each other. The alloy exhibits flow instability at lower temperatures and higher strain rates, which manifests as adiabatic shear bands. A comparison of the processing maps of these alloys revealed that the hot deformation mechanisms have not changed significantly by the increase of Ca addition.

Keywords: thermomechanical processing; magnesium alloy; calcium addition; disintegrated melt deposition; processing map; formability

---

1. Introduction

In recent years, magnesium alloys have become popular for use as lightweight structural components in automobile, aerospace and biomedical applications [1,2]. AZ31 is the most popular wrought magnesium alloy that has been investigated extensively [1,3]. However, the alloy faces limitations of creep strength and corrosion resistance. Its creep strength may be enhanced by the addition of alloying additions like rare-earth or alkaline-earth elements. The addition of Ca to magnesium alloys has been studied extensively in view of its cost advantage over the rare-earth elements [4–6] and its effect on strength and hot workability has been reviewed recently [6]. The room temperature formability of AZ31 is improved by the addition of 0.5% Ca [7,8] since it reduces the basal texture promoting prismatic slip. Sakai et al. [9] reported that addition of 1%Ca enhances the room temperature strength of AZ31 significantly although the ductility improves only at temperatures higher than about 150 °C. Ca addition to an extent of 0.7% enhances the corrosion resistance of AZ31 alloy [10] although higher additions are not particularly beneficial. The ultimate compressive strength of AZ31 alloy at its service temperature (about 150 °C) increases from 134 MPa to 235 MPa with 2% Ca addition [6]. Further additions of Ca to AZ31 hold a potential in improving its higher temperature strength and in this study, it is proposed to explore this possibility by enhancing the Ca content to 3%.

Complex intermetallic phases of $(Mg,Al)_2Ca$ and $Ca_2Mg_6Zn_3$ are formed when alloying elements of Al, Zn, and Ca are added to magnesium, and they become dispersed in the microstructure with different morphologies depending on the melting and casting processes adopted [4,11,12]. However, a technique called disintegrated melt deposition (DMD) has been developed to inherit the combined benefits of stir-casting and spray-processing techniques [13,14]. Similar to stir casting, this method involves the vortex mixing of alloying elements/reinforcement in the molten matrix and the resulting slurry is deposited onto a metallic substrate after disintegration by jets of inert gases. Unlike conventional spray processes, this method employs lower impinging velocity of gas jets with almost 100% recovery of poured material as the formation of overspray powders is avoided [13]. Therefore, the DMD technique offers the features of: (i) simplicity and cost effectiveness of conventional stir cast foundry process and (ii) fine grain and homogeneous microstructure associated with spray deposition process.

DMD technique of processing involves atomization of liquid metal alloy which helps in the production and deposition of fine metal droplets to form fine grains and a uniform distribution of intermetallic phases in the microstructure. The collected metal droplets form a billet of low density and can be further consolidated by extruding at desired temperature to produce rods/bars of desired size [15,16]. The so-called DMD rods of AZ31, AZ31−1Ca and AZ31-2Ca alloys prepared by DMD route have been studied earlier as regards their microstructure and hot working mechanisms [17–19]. In these investigations, it is found that addition of Ca to AZ31 has enhanced its compressive strength without restricting the hot workability. The aim of the investigation is to examine the effect of further addition of Ca to an extent of 3% on the hot workability of AZ31. In order to introduce a higher Ca content in the alloy and distribute it homogeneously, the advantage of the DMD processing route has been exploited. As in the case of earlier investigations of hot workability of DMD AZ31 alloys, the approach of processing maps has been adopted in this study also.

2. Methodology

The principles behind the development of processing maps have been extensively documented in the literature [20–22] following a dynamic material modeling approach. In simple terms, the model differentiates the way the applied energy to deform a material is utilized or dissipated. The relative proportion of energy dissipated for the generation of heat and microstructural changes determines the efficiency of power dissipation. In the latter case, the energy dissipation may result in increase of dislocations, dynamic recovery and recrystallization, etc. A simple parameter that can provide an estimate of power partition between heat and microstructural changes in the strain rate sensitivity (m) of flow stress under a given set of deforming conditions of temperature and strain rate. The efficiency of power dissipation (η), responsible for bringing the microstructural changes during deformation can be calculated using:

$$\eta = 2m/(m + 1) \tag{1}$$

It is convenient to present the dissipation efficiency in the form of a map consisting of iso-efficiency contours over the chosen ranges of temperature and strain rate to distinguish their effects.

Another important consideration is whether the material flow is stable during deformation under the chosen forming conditions. In simple terms, the material flow rate should not exceed the applied velocity. For example, fracture or wedge cracking may propagate at a faster rate than the moving tool that imparts energy, which is not desirable in forming processes. A stable flow in general enables material integrity after hot deformation and results in better product. The transition from stable to unstable flow can be estimated using the extremum principles of irreversible thermodynamics as applied to continuum mechanics of large plastic flow [23]. The criterion for the onset of flow instability is:

$$\xi(\dot{\varepsilon}) = \frac{\partial \ln[m/(m+1)]}{\partial \ln \dot{\varepsilon}} + m \le 0 \tag{2}$$

where ξ is the instability parameter for a given strain rate ($\dot{\varepsilon}$). The flow becomes localized and causes flow instability for deformation conditions where $\xi(\dot{\varepsilon}) \leq 0$.

A processing map integrates both the important considerations mentioned above, namely, dissipation efficiency and flow stability, by superimposing their contour maps to provide a comprehensive understanding of material response under applied process conditions of temperature and strain rate. High dissipation efficiency domains that form hills in contour or 3-dimensional maps with stable flow can be identified easily where the workability of a material is optimal due to the occurrence of favorable mechanisms, such as dynamic recrystallization (DRX). Similarly, conditions of flow instability, even though dissipation efficiency may be high, can be earmarked as failure regimes that should be avoided for hot deformation of the material. It is possible that several desirable domains and undesirable regimes may be found over the chosen ranges of temperature and strain rate due to specific deformation mechanisms that are operative.

The commonly used kinetic rate equation relating the flow stress (σ) to strain rate ($\dot{\varepsilon}$) and temperature (T) of hot deformation is given by Jonas et al., [24]:

$$\dot{\varepsilon} = A\sigma^{n}\exp[-Q / RT] \tag{3}$$

where A is a constant, n is a stress exponent, Q is the activation energy, and R is the gas constant. This kinetic rate equation is obeyed within a domain where single deformation mechanism dominates and, therefore, can be applied to evaluate the associated apparent activation energy. The activation parameters n and Q facilitate the identification of rate-controlling deformation mechanisms in corresponding domains.

3. Materials and Methods

The AZ31–3 wt% calcium (AZ31-3Ca) alloy was prepared by the disintegrated melt deposition (DMD) processing technique developed by Gupta and coworkers [15,16]. Briefly, the procedure consisted of the following two steps: (1) Primary processing: Rectangular pieces of AZ31 magnesium alloy were cut from ingots. Holes of 12 mm diameter and 30 mm depth were drilled in these rectangular pieces and were filled with the required amount of calcium. They were melted in a crucible that has bottom pouring arrangement and heated to reach a temperature of 750 °C. The molten alloy was stirred for 5 min at 450 rev min^{-1} using a twin blade (pitch 45°) mild steel impeller to facilitate chemical homogenization. The impeller was coated with Zirtex 25 (86% ZrO_2, 8.8% Y_2O_3, 3.6% SiO_2, 1.2% K_2O and Na_2O, and 0.3% trace inorganic) to avoid iron contamination. The melt was then released through a 10 mm diameter orifice at the base of the crucible and the droplets of the melt were disintegrated by two jets of argon gas orientated normal to the melt stream. The argon gas flow rate was maintained at 25 L min^{-1}. The disintegrated melt was deposited onto stainless steel substrate to obtain a preform billet of 40 mm diameter. (2) Secondary processing: The preformed billet was machined to 35 mm diameter, heated to 400 °C and kept for 1 h before directly extruding through a preheated die at 350 °C with colloidal graphite used as a lubricant. Rods of 9.8 mm diameter were obtained using an extrusion ratio of about 13 adopting this direct extrusion procedure.

The experimental procedure for hot compression testing was described in detail earlier [25]. Briefly, cylindrical specimens of 9.8 mm in diameter and 15 mm in height were machined from the extruded rods for uniaxial compression along the direction of extrusion. A hole of 1 mm diameter was machined at mid-height to reach the center of the specimen. A thermocouple was inserted in the hole to monitor the specimen temperature and also to measure the instantaneous temperature rise during deformation. The adiabatic temperature rise was measurable for tests conducted at strain rates of 0.1 s^{-1} and higher. Compression tests were carried out at strain rates in the range of 0.0003–10 s^{-1} and temperatures in the range of 300–450 °C. The selected constant true strain rate during a compression test was obtained through an exponential decay in the actuator speed of the machine. Graphite powder mixed with grease was applied as the lubricant at the specimen-die interfaces in all the experiments. The specimens were deformed up to a true strain of about 1 and quickly cooled in water. The load-stroke data were converted into true stress-true strain values using standard equations. The true stress values were plotted as a function of measured temperature at different strain levels and were fitted by smooth curves. The corrected flow stress was obtained from these fitted curves at any selected temperature for calculating the strain rate sensitivity of flow stress. The processing map was developed at different strains from the variations of strain rate sensitivity of flow stress with strain rate at different temperatures using Equations (1) and (2) and the procedure described earlier [20]. The deformed specimens were sectioned in the center, parallel to the compression axis, and were mounted, polished and etched with acetic picral solution for metallographic examination.

4. Results and Discussion

The starting microstructure of DMD processed AZ31-3Ca alloy is shown in Figure 1a,b as viewed in an optical microscope and in scanning electron microscope, respectively. The grain size is fine with the average grain diameter in the range 2–3 μm, as determined using linear intercept method. The microstructure has a large volume fraction of $(Mg,Al)_2Ca$ intermetallic particles distributed uniformly in the microstructure along with a random distribution of a small volume fraction of Al_8Mn_5 particles.

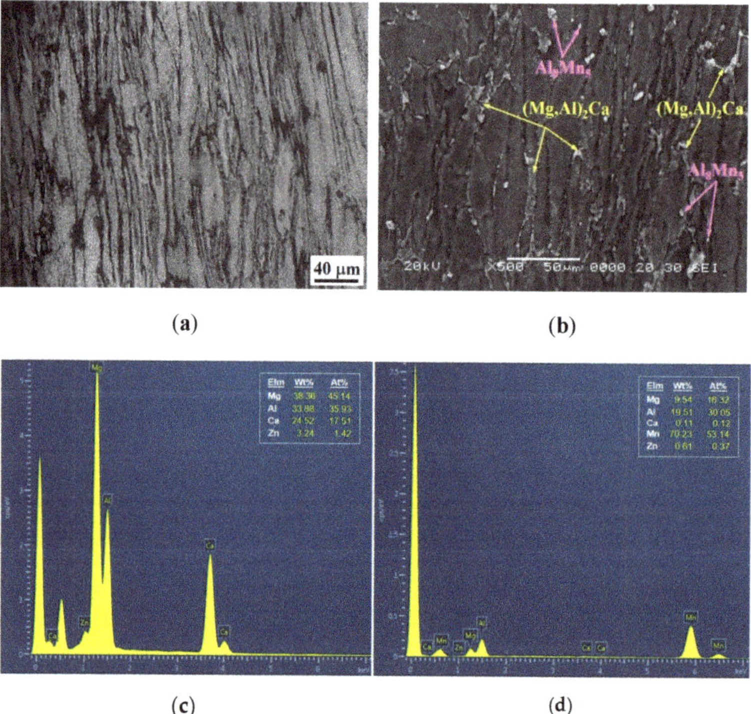

Figure 1. Starting microstructure of disintegrated melt deposition (DMD)-processed AZ31-3Ca. (**a**) Optical, (**b**) scanning electron microscope (SEM) micrograph, and (**c**,**d**) energy dispersive spectroscopy (EDS) spectra at locations of intermetallic particles.

The elemental distribution of Al, Ca, Zn and Mn as obtained by energy dispersive spectroscopy (EDS) mapping is shown in Figure 2 which confirms the uniformity of their distribution. The formation of the above two phases is according to the predictions of thermodynamic model [26,27], and comparable to those reported in the literature [6,12,28]. The elemental composition at the selected spots on the intermetallic particles/phases is revealed from EDS spectra shown Figure 1c,d. These spectra indicated high concentrations of Mg/Al and Ca in the case of $(Mg,Al)_2Ca$ particles, and Al and Mn in the case of Al_8Mn_5. Considering the influence of surrounding material around these fine intermetallic particles, it may be noted that compositional analysis obtainable from EDS is not an accurate representation of that of the particles but assist in identifying them with the support of thermodynamic prediction model mentioned above.

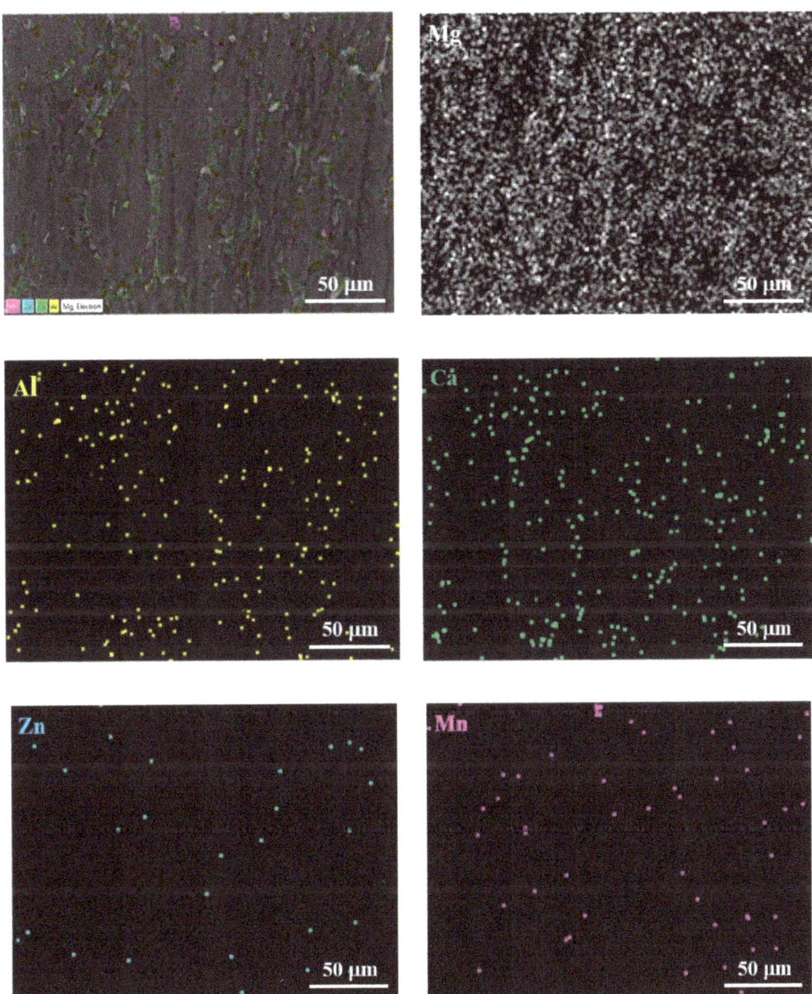

Figure 2. EDS (energy dispersive spectroscopy) elemental mapping for AZ31-3Ca DMD alloy.

The compressive strength of the AZ31-3Ca alloy measured at a temperature 150 °C, typical heat resistant limit for magnesium alloys, along with the strengths of AZ31 alloy, AZ31-1Ca, and AZ31-2Ca are shown in Figure 3. It can be seen that the ultimate compressive strength (UCS) has significantly increased with the addition of Ca, with UCS of AZ31-3Ca nearly double that of base AZ31 alloy. However, AZ31-3Ca exhibited highest brittleness as well that makes forming most difficult, vindicating the need to establish precise thermomechanical processing options.

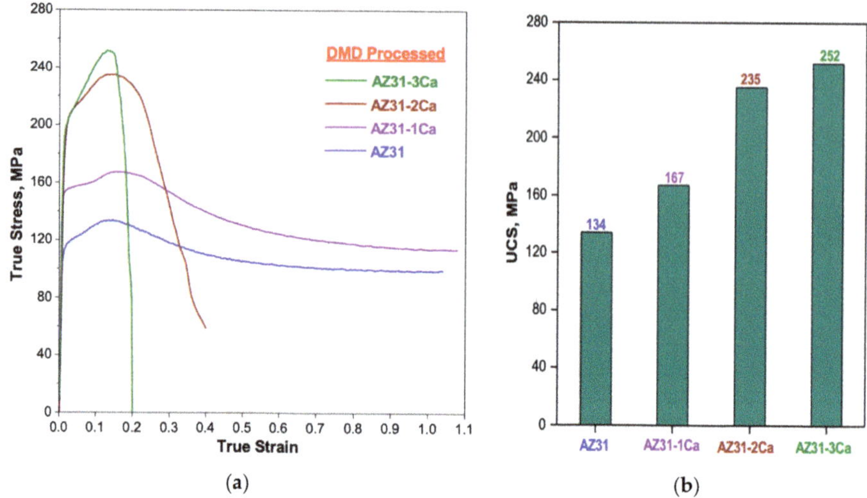

(a) (b)

Figure 3. (**a**,**b**) Stress-strain curves obtained in compression for AZ31, AZ31-1Ca, AZ31-2Ca, and AZ31-3Ca alloys. The ultimate compression strengths of the alloys are also shown for comparison.

The true stress–true strain curves obtained in compression of DMD processed AZ31-3Ca alloy at 300 °C and 400 °C and at different strain rates are shown in Figure 4a,b, respectively. The curves at 300 °C exhibited flow softening at strain rates higher than about 0.01 s^{-1} and are of steady-state type at lower strain rates. However, at higher temperatures the steady-state flow occurred at higher strain rates. The shapes of deformed specimens are shown in Figure 5 and the alloy exhibited homogeneous flow at all temperatures and strain rates.

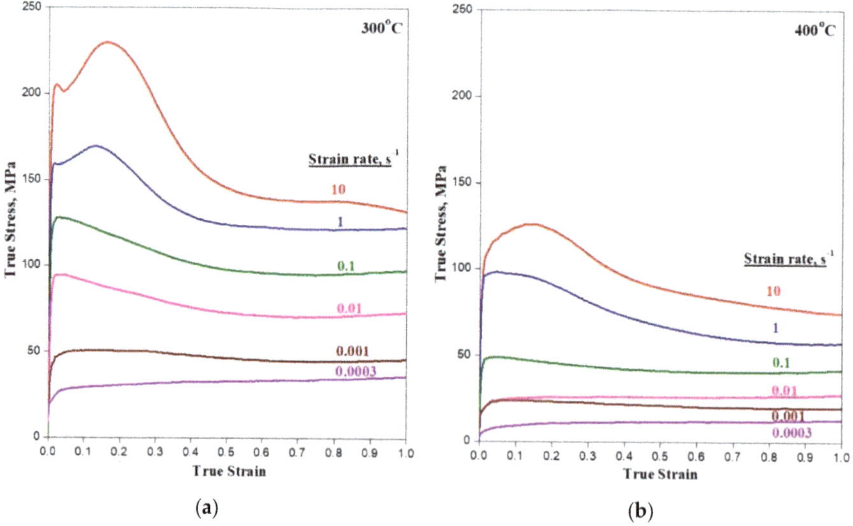

(a) (b)

Figure 4. True stress–true strain curves obtained in compression on DMD processed AZ31-3Ca alloy at different strain rates at temperatures of (**a**) 300 °C, and (**b**) 400 °C.

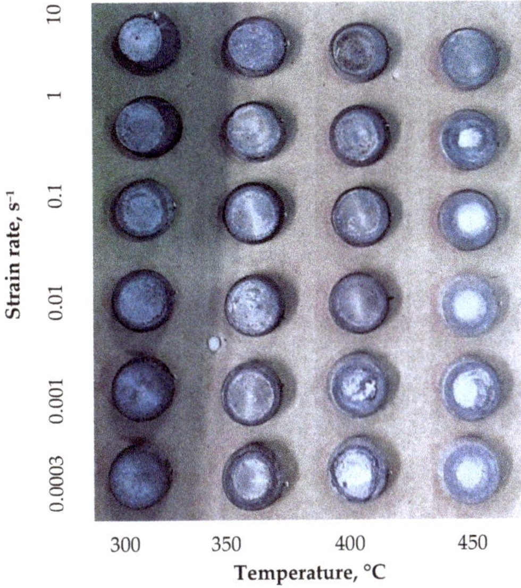

Figure 5. Shapes of deformed cylindrical specimens of DMD processed AZ31-3Ca alloy compressed at different temperatures and strain rates.

The processing map obtained at a true strain of 0.7 is shown in Figure 6. The numbers assigned to the contours represent efficiency of power dissipation expressed in percent. The maps at other strains are not significantly different after steady-state flow has occurred. The map exhibits three domains in the temperature and strain rate ranges given below.

Domain (1): 300–375 °C and 0.0003–0.003 s^{-1} with a peak efficiency of about 53% at 375 °C/0.0003 s^{-1}.
Domain (2): 375–450 °C and 0.0003–0.003 s^{-1} with a peak efficiency of about 58% at 405 °C/0.0003 s^{-1}.
Domain (3): 375–450 °C and 0.1–10 s^{-1} with a peak efficiency of about 35% at 425 °C/0.1 s^{-1}.

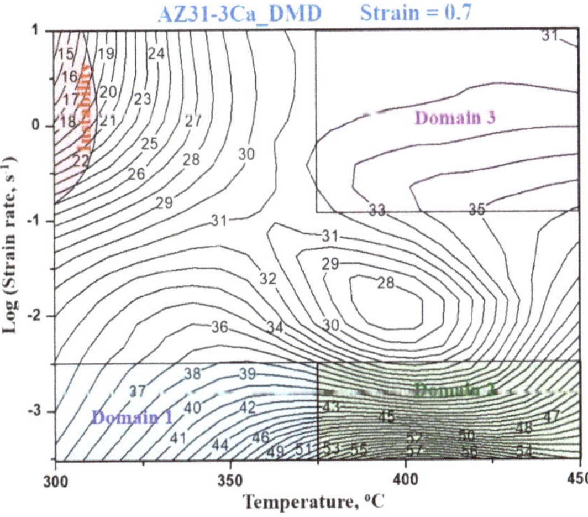

Figure 6. Processing map for DMD processed AZ31-3Ca alloy at a strain of 0.7. The numbers on the contours represent efficiency of power dissipation in percent.

Domains (1) and (2) have merged at about 375 °C in view of the higher efficiency range exhibited by Domain (2).

The microstructure obtained on a specimen deformed at 300 °C/0.0003 s^{-1} in Domain (1) is shown in Figure 7a. It exhibits fine grain size suggesting that dynamic recrystallization occurs in this domain. At temperatures corresponding to this domain, basal slip along with prismatic slip occurs. In view of the slower strain rates, the recovery mechanism that nucleates DRX will be climb of edge dislocations involving lattice self-diffusion, although limited amount of cross-slip may be associated with prismatic slip. Kinetic analysis of the temperature and strain rate dependence of flow stress has been conducted using Equation (3). The variation of normalized flow stress with logarithms of strain rate is shown in Figure 8a and the Arrhenius plot showing the variation of normalized flow stress with inverse of absolute temperature is shown in Figure 8b. It may be noted that the kinetic rate equation is obeyed within the deterministic domains. The stress exponent obtained in Domain (1) is 4.29 and the apparent activation energy is 158 kJ/mole. For the mechanism of climb, which is controlled by lattice self-diffusion, the activation energy is expected to be 135 kJ/mole [29]. The estimated value is higher than that for lattice self-diffusion, suggesting that the high-volume fraction of Ca-containing intermetallic particles causes considerable back stress enhancing the apparent activation energy.

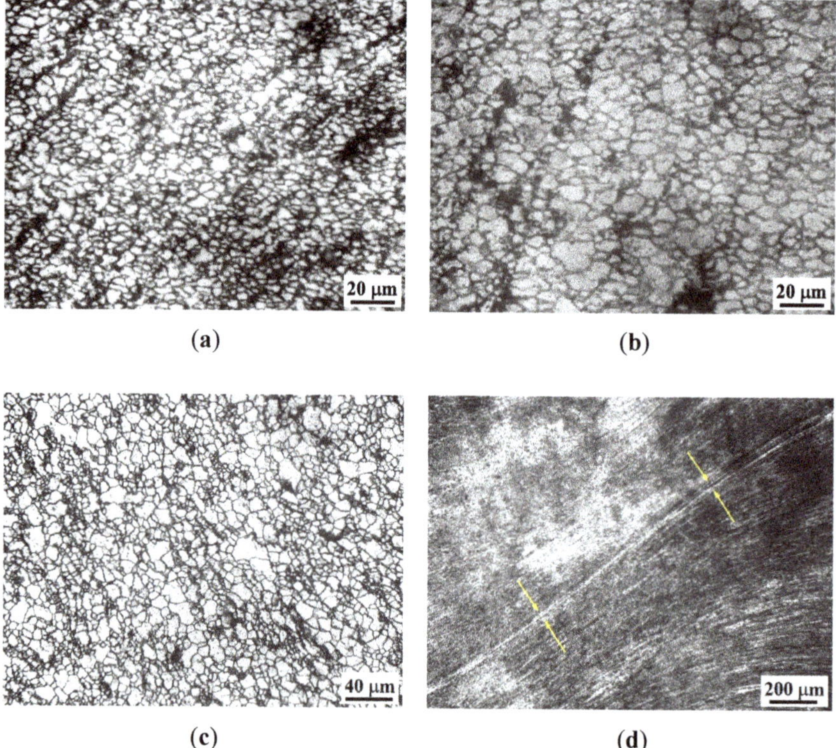

(a) (b) (c) (d)

Figure 7. Microstructures obtained on DMD processed AZ31-3Ca alloy deformed at (**a**) 350 °C/ 0.0003 s^{-1} (Domain 1) (**b**) 400 °C/0.0003 s^{-1} (Domain 2) (**c**) 450 °C/10 s^{-1} (Domain 3), and (**d**) 300 °C/ 10 s^{-1} (instability regime). The compression axis is vertical.

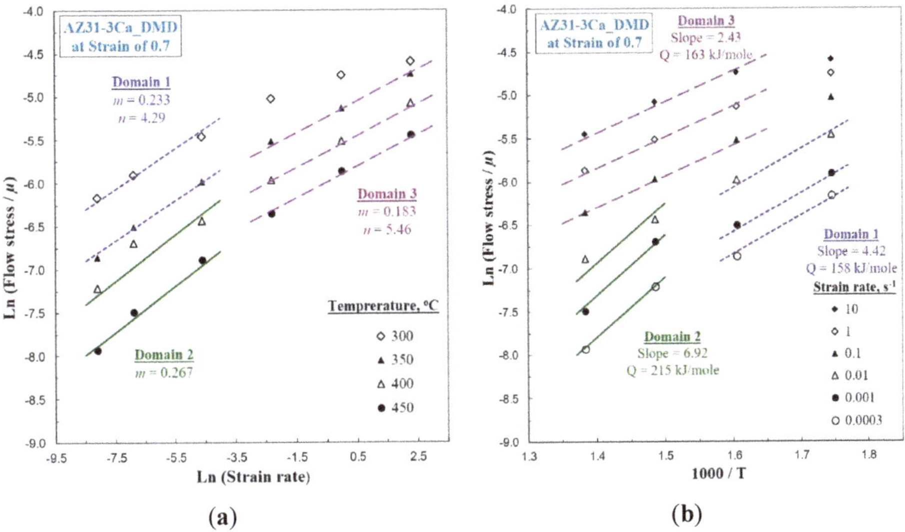

Figure 8. (**a**) Variation of normalized flow stress with strain rate on natural logarithmic scale, and (**b**) Arrhenius plot showing variation of normalized flow stress with inverse of absolute temperature for DMD processed AZ31-3Ca alloy.

The microstructure obtained on a specimen deformed at 400 °C and 0.0003 s^{-1} near the peak efficiency in Domain (2) is shown in Figure 7b which exhibits equiaxed dynamically recrystallized structure. In addition, intercrystalline cracking has occurred during deformation. Firstly, in view of the higher temperature range in which this domain occurs, second-order pyramidal slip occurs readily. The availability of large number of intersecting slip systems and the higher stacking fault energy (172 mJ/m^2) [30] promote the occurrence of extensive cross-slip which nucleates DRX. Secondly, once the fine-grained structure is produced by DRX, grain boundary sliding occurs since the strain rates are slow resulting in wedge cracking and cracking along the grain boundaries. The high efficiency in this domain (58%) and the steep increase in efficiency with decreasing strain rate (appearing as closely space contours representing steep efficiency hill) are also indicative of such a grain boundary sliding mechanism. Thus, the end result of deformation in this domain is grain boundary cracking, which is not desirable.

The microstructure of a specimen deformed at 450 °C/10 s^{-1} (Domain 3) is shown in Figure 7c and exhibits DRX. Within the temperature range in this domain (375–450 °C), basal, prismatic as well as second-order pyramidal slip systems can cause deformation. However, since this domain occurs at higher strain rates (0.1–10 s^{-1}), the recovery mechanism is likely to be grain boundary self-diffusion since lattice self-diffusion is a slower process and higher strain rates slow down recovery by cross-slip. The apparent activation energy estimated in this domain (Figure 8) is 163 kJ/mole, which is higher than the activation energy for grain boundary self-diffusion in Mg (95 kJ/mole) [30]. The apparently higher value may be attributed to the back-stress caused by the large volume fraction of Ca-containing intermetallic particles in the microstructure.

The processing map exhibits a regime of flow instability in the temperature range 300–310 °C at strain rates higher than about 0.1 s^{-1}. The microstructure of the specimen deformed at 300 °C/10 s^{-1} is shown in Figure 7d which reveals that the flow instability manifestation is adiabatic shear band formation (marked by arrows). Cracking has occurred along the shear band due to high intensity of localization which is nearly at 45° with respect to the compression axis.

An interesting feature exhibited by the processing map is the occurrence of a bifurcation in the temperature range 325–425 °C and strain rate range 0.003–0.1 s^{-1}, where the efficiency has reached a low value of 28%. This represents a region between the three domains identified above. In dynamic systems terminology [31], the bifurcation is akin to a dissipative energy hill and may be viewed as a saddle configuration. Processing under conditions of bifurcation will lead to microstructural instability since any small perturbation will trigger a different mechanism surrounding the bifurcation. The microstructures of the specimen deformed at 400 °C and at two strain rates (0.001 and 0.01 s^{-1}) near the bifurcation region are shown in Figure 9a,b. The microstructural instability is manifested in terms of a combination of large and small grains in Figure 9a and very fine grains in Figure 9b, revealing that microstructural control is difficult if processed around the bifurcation region.

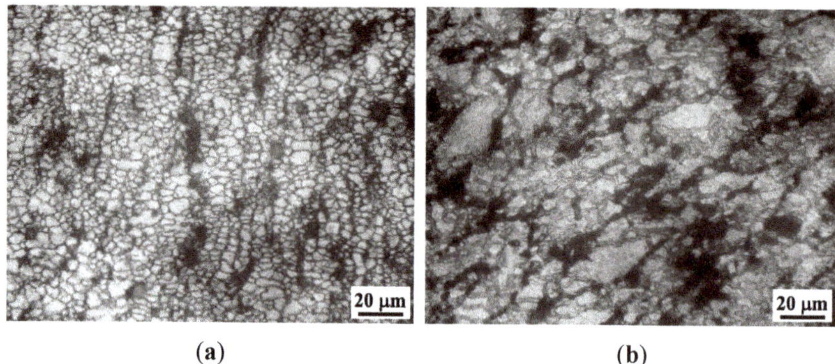

Figure 9. Microstructures obtained on DMD processed AZ31-3Ca alloy specimens deformed at (**a**) 400 °C/0.001 s^{-1} (**b**) 400 °C/0.01 s^{-1} (near bifurcation region).

The processing maps for DMD processed AZ31 [17], AZ31-1Ca [18], AZ31-2Ca [19] are compared in Figure 10 with that obtained currently on AZ31-3Ca alloy at a strain of 0.7. The starting microstructure, the characteristics of the different domains exhibited by the maps and the interpretations are shown in Table 1. With increasing Ca content in AZ31, the volume fraction of Ca-containing intermetallic phase $(Mg,Al)_2Ca$ increases and the grain size decreases to about 2–3 µm. The increasing volume fraction of the intermetallic phase has an effect on the hot working behavior of AZ31.

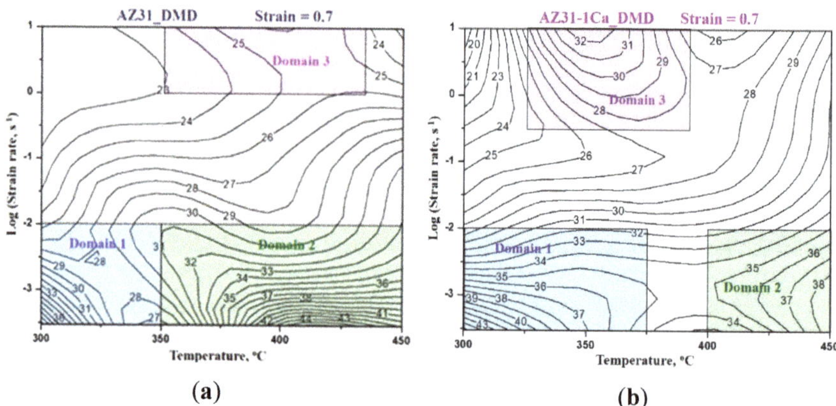

Figure 10. *Cont.*

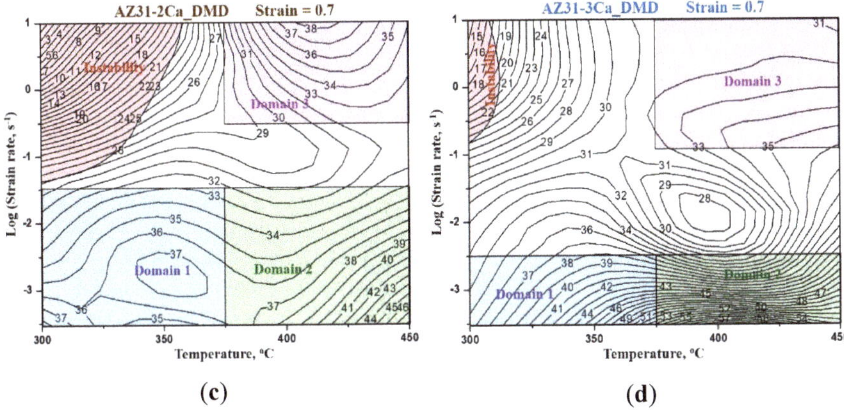

Figure 10. Processing maps obtained at a strain of 0.7 on DMD processed (**a**) AZ31, (**b**) AZ31-1Ca, (**c**) AZ31-2Ca, and (**d**) AZ31-3Ca alloys. The numbers indicated on the contours represent efficiency of power dissipation expressed as percent.

Table 1. Details of (**a**,**b**) microstructural parameters, (**c**) temperature—strain rate ranges for domains, (**d**) peak efficiency (η) conditions, (**e**) apparent activation energy (Q_{app}) and (**f**) proposed mechanisms (DRX: dynamic recrystallization) in the various domains of processing maps for DMD processed AZ31, AZ31-1Ca, AZ31-2Ca and AZ31-3Ca magnesium alloys.

Parameter		AZ31 (Ref. 17)	AZ31-1Ca (Ref. 18)	AZ31-2Ca (Ref. 19)	AZ31-3Ca (current work)
Initial grain size, µm	(a)	9	3	3	2–3
Phases in the microstructure	(b)	$Mg_{17}Al_{12}$	$(Mg,Al)_2Ca$ Al_8Mn_5	$(Mg,Al)_2Ca$ Al_8Mn_5	$(Mg,Al)_2Ca$ Al_8Mn_5
Domain (1)	(c)	300–350 °C 0.0003–0.01 s^{-1}	300–375 °C 0.0003–0.01 s^{-1}	300–375 °C 0.0003–0.03 s^{-1}	300–375 °C 0.0003–0.003 s^{-1}
	(d)	300 °C/0.0003 s^{-1}; peak η: 39%	300 °C/0.0003 s^{-1}; peak η: 45%	350 °C/0.001 s^{-1} peak η: 37%	375 °C/0.0003 s^{-1}; peak η: 53%
	(e)	Q_{app}: 126 kJ/mole	Q_{app}: 112 kJ/mole	Q_{app}: 133 kJ/mole	Q_{app}: 158 kJ/mole
	(f)	DRX: Basal slip + climb by Lattice self-diffusion	DRX: Basal slip + climb by lattice self-diffusion	DRX: prismatic slip + climb by lattice self-diffusion	DRX: Basal slip + prismatic slip + climb by lattice self-diffusion
Domain (2)	(c)	350–450 °C 0.0003–0.01 s^{-1}	400 – 450 °C 0.0003–0.01 s^{-1}	375–450 °C 0.0003–0.03 s^{-1}	375–450 °C 0.0003–0.003 s^{-1}
	(d)	425 °C/0.0003 s^{-1}; peak η: 46%	450 °C/0.0003 s^{-1}; peak η: 38%	450 °C/0.0003 s^{-1}; peak η: 48%	405 °C/0.0003 s^{-1}; peak η: 58%
	(e)	Q_{app}: 144 kJ/mole	Q_{app}: 166 kJ/mole	Q_{app}: 192 kJ/mole	Q_{app}: 215 kJ/mole
	(f)	Wedge cracking	Wedge cracking	Second-order pyramidal slip +cross slip	Grain boundary cracking
Domain (3)	(c)	350–435 °C 1–10 s^{-1}	325–390 °C 0.3–10 s^{-1}	375–450 °C 0.3–10 s^{-1}	375–450 °C 0.1–10 s^{-1}
	(d)	400 °C/10 s^{-1}, peak η: 26%	350 °C/10 s^{-1}, peak η: 32%	400 °C/10 s^{-1}, peak η: 38%	425 °C/0.1 s^{-1}; peak η: 35%
	(e)	Q_{app}: 127 kJ/mole	Q_{app}: 87 kJ/mole	Q_{app}: 166 kJ/mole	Q_{app}: 163 kJ/mole
	(f)	DRX: climb by grain boundary self-diffusion	DRX: climb by grain boundary self-diffusion	DRX: climb by grain boundary self-diffusion	DRX: climb by grain boundary self-diffusion

The characteristics of Domain (1) did not change significantly although the temperature for peak efficiency and the apparent activation energy have increased with increasing Ca content. In this domain, DRX occurs by basal slip + prismatic slip and recovery primarily by climb of edge dislocations controlled by lattice self-diffusion. The higher back-stress generated by the presence of intermetallic particles increases the back-stress to dislocation generation as well as dynamic recovery process which is responsible for moving the domain to higher temperatures and for higher apparent activation energy.

Domain (2) occurs in similar temperature range in all the alloys although the temperature at which peak efficiency occurs increases in AZ31-1Ca and AZ31-2Ca alloys and decreases in AZ31-3Ca alloy. The peak efficiency exhibits a high value in AZ31-3Ca alloy and the apparent activation energy is higher than that for lattice self-diffusion in all the alloys. In view of the higher temperature in which this occurs, DRX occurs by second-order pyramidal slip and recovery by cross-slip. The fine-grained structure produced by this DRX process promotes grain boundary sliding during large deformation, which results in wedge cracking and intercrystalline fracture. However, in AZ31-2Ca alloy, wedge cracking did not occur since the grain boundary sliding process is restricted by Ca-containing intermetallic particles. When the Ca content is further increased to 3%, the grain boundaries are weakened by a large volume content of the particles resulting in intercrystalline cracking. Thus, processing in Domain (2) which occurs at higher temperatures and lower strain rates, is not desirable since it results in intercrystalline cracking except in AZ31-2Ca alloy.

Domain (3) moves to higher temperatures with increasing Ca content in AZ31. In this domain, non-basal slip systems operate extensively to cause DRX. The recovery mechanisms of climb by lattice diffusion and cross-slip to nucleate DRX are restricted in the high strain rate range at which this domain occurs. Instead, recovery occurs by grain boundary self-diffusion. The apparent activation energy values are higher than that expected for grain boundary self-diffusion (95 kJ/mole) [29] which may be attributed to the higher back stress caused by the Ca-containing intermetallic particles. It may be noted that much slower rates are required for recovery in AZ31-3Ca since a larger volume content of Ca-containing intermetallic particles will be present at the grain boundaries slowing the rate of recovery by grain boundary self-diffusion. In all the Ca-containing AZ31 DMD processed alloys, hot working is probably best done in this domain.

5. Conclusions

The hot deformation mechanisms in DMD processed AZ31-3Ca alloy have been evaluated using processing maps, and the behavior has been compared with that of base AZ31, AZ31-1Ca and AZ31-2Ca alloys with a view to bringing out the effect of Ca addition. The following conclusions are drawn from this investigation.

(1) Addition of 3% Ca to AZ31 refines the grain size and increases the volume fraction of $(Mg,Al)_2Ca$ intermetallic phase, which is distributed uniformly in the microstructure.
(2) The processing map for AZ31-3Ca alloy exhibited three domains in the temperature strain rate ranges: (1) 300–375 °C and 0.0003–0.003 s^{-1}, (2) 375–450 °C and 0.0003–0.003 s^{-1}, and (3) 375–450 °C and 0.1–10 s^{-1}.
(3) In Domain (1), DRX occurs when basal + prismatic slip associated with dynamic recovery by climb controlled by lattice self-diffusion. In Domain (2), the plastic flow results in intercrystalline cracking promoted by grain boundary sliding initiating wedge cracks. In Domain (3) that occurs at higher strain rates, DRX occurs by non-basal slip with recovery by climb controlled by grain boundary self-diffusion. This recovery is promoted by fine grain size in the alloy.
(4) The alloy exhibits some flow instability at lower temperatures and higher strain rates, and this manifests as adiabatic shear bands.
(5) A comparison of processing maps on the Ca containing AZ31 alloys revealed that the hot deformation mechanisms have not significantly changed by increasing Ca addition except in

AZ31-2Ca alloy where wedge cracking is avoided in Domain (2) due to the reduction of grain boundary sliding that is attributed to pinning by intermetallic particles.

Author Contributions: K.P.R. was responsible for conceptualization, funding acquisition, and supervision of the work, as well as performed analysis of the data and writing the paper; K.S. performed the experimental work, generating the processing maps and microstructural work; Y.V.R.K.P. contributed towards analysis of the data, processing maps, and writing the paper; M.G. contributed towards developing the DMD processed alloys. All authors have read and agreed to the published version of the manuscript.

Funding: This research was funded by a Strategic Research Grant (Project #7002704) from the City University of Hong Kong and a General Research Fund (Project #114811) from the Research Grants Council of Hong Kong SAR, China.

Acknowledgments: The authors would like to thank Yuen Chi Hung, Technical Officer, Department of Biomedical Engineering, City University of Hong Kong, for conducting all the hot compression tests involved in this research work. The help received from C. Dharmendra, Marine Additive Manufacturing Centre of Excellence (MAMCE), University of New Brunswick, Fredericton, E3B5A1, NB, Canada, (formerly with City University of Hong Kong), for some of the studies involved in this research is acknowledged.

Conflicts of Interest: The authors declare no conflict of interest. The funders had no role in the design of the study; in the collection, analyses, or interpretation of data; in the writing of the manuscript; or in the decision to publish the results.

References

1. Kainer, K.U.; Dieringa, H.; Dietzel, W.; Hort, N.; Blawert, C. The use of magnesium alloys: Past, present and future. In *Magnesium Technology in Global Age*; Pekguleryuz, M.O., Mackenzie, L.W.F., Eds.; Canadian Institute of Mining, Metallurgy and Petroleum: Montreal, QC, Canada, 2006; pp. 3–19.
2. Schilling, T.; Bauer, M.; Hartung, D.; Brandes, G.; Tudorache, I.; Cebotari, S.; Meyer, T.; Wacker, F.; Haverich, A.; Hassel, T. Stabilisation of a segment of autologous vascularised stomach as a patch for myocardial reconstruction with degradable magnesium alloy scaffolds in a swine model. *Crystals* **2020**, *10*, 438. [CrossRef]
3. Vespa, G.; Mackenzie, L.W.F.; Verma, R.; Zarandi, F.; Essadiqi, E.; Yue, S. The influence of the as-hot rolled microstructure on the elevated temperature mechanical properties of magnesium AZ31 sheet. *Mater. Sci. Eng. A* **2008**, *487*, 243–250. [CrossRef]
4. Masoudpanah, S.M.; Mahmudi, R. Effects of rare-earth elements and Ca additions on the microstructure and mechanical properties of AZ31 magnesium alloy processed by ECAP. *Mater. Sci. Eng. A* **2009**, *526*, 22–30. [CrossRef]
5. Chaudry, U.M.; Hamad, K.; Kim, J.-G. Ca-induced plasticity in magnesium alloy: EBSD measurements and VPSC calculations. *Crystals* **2020**, *10*, 67. [CrossRef]
6. Rao, K.P.; Prasad, Y.V.R.K.; Dharmendra, C.; Suresh, K.; Hort, N.; Dieringa, H. Review on hot working behavior and strength of calcium-containing magnesium alloys. *Adv. Eng. Mat.* **2018**, *20*, 1701102. [CrossRef]
7. Chaudry, U.M.; Kim, T.H.; Park, S.D.; Kim, Y.S.; Hamad, K.; Kim, J.-G. On the high formability of AZ31-0.5Ca magnesium alloy. *Materials* **2018**, *11*, 2201. [CrossRef]
8. Chaudry, U.M.; Kim, Y.S.; Hamad, K. Effect of Ca addition on the room temperature formability of AZ31 magnesium alloy. *Mater. Lett.* **2019**, *238*, 305–308. [CrossRef]
9. Sakai, N.; Funami, K.; Noda, M.; Mori, H. Effect of Ca addition on the high temperature deformation behavior of AZ31 magnesium alloy. In *Proceedings of the 8th Pacific Rim International Congress on Advanced Materials and Processing*; Marquis, F., Ed.; Springer: Cham, Switzerland, 2013; pp. 1203–1209.
10. Yim, C.D.; Kim, Y.M.; You, B.S. Effect of Ca addition on the corrosion resistance of gravity cast AZ31 magnesium alloy. *Mater. Trans.* **2007**, *48*, 1023–1028. [CrossRef]
11. Jeong, J.; Im, J.; Song, K.; Kwon, M.; Kim, S.K.; Kang, Y.-B.; Oh, S.H. Transmission electron microscopy and thermodynamic studies of CaO-added AZ31 Mg alloys. *Acta Mater.* **2013**, *61*, 3267–3277. [CrossRef]
12. Kim, Y.M.; Yim, C.D.; You, B.S. Effect of the Ca content on the microstructural evolution of Ca containing AZ31 cast alloys. *Met. Mater. Int.* **2011**, *17*, 583–586. [CrossRef]
13. Gupta, M.; Lai, M.O.; Soo, C.Y. Processing-microstructure-mechanical properties of an Al-Cu/SiC metal matrix composite synthesized using disintegrated melt deposition technique. *Mater. Res. Bull.* **1995**, *30*, 1525–1534. [CrossRef]

14. Gupta, M.; Sharon, N.M.L. *Magnesium, Magnesium Alloys, and Magnesium Composites*; John Wiley & Sons: Hoboken, NJ, USA, 2011; ISBN 978-0-470-49417-2.
15. Hassan, S.F.; Gupta, M. Development of nano-Y_2O_3 containing magnesium nanocomposites using solidification processing. *J. Alloy. Compd.* **2007**, *429*, 176–183. [CrossRef]
16. Nguyen, Q.B.; Gupta, M. Improving compressive strength and oxidation resistance of AZ31B magnesium alloy by addition of nano-Al_2O_3 particulates and Ca. *J. Compos. Mater.* **2010**, *44*, 883–896. [CrossRef]
17. Zhong, T.; Rao, K.P.; Prasad, Y.V.R.K.; Gupta, M. Processing maps, microstructure evolution and deformation mechanisms of extruded AZ31-DMD during hot uniaxial compression. *Mater. Sci. Eng. A* **2013**, *559*, 773–781. [CrossRef]
18. Rao, K.P.; Zhong, T.; Prasad, Y.V.R.K.; Suresh, K.; Gupta, M. Hot working mechanisms in DMD-processed versus cast AZ31-1 wt.% Ca alloy. *Mater. Sci. Eng. A* **2015**, *644*, 184–193. [CrossRef]
19. Rao, K.P.; Suresh, K.; Prasad, Y.V.R.K.; Hort, N.; Gupta, M. Enhancement of strength and hot workability of AZX312 magnesium alloy by disintegrated melt deposition (DMD) processing in contrast to permanent mold casting. *Metals* **2018**, *8*, 437. [CrossRef]
20. Prasad, Y.V.R.K.; Rao, K.P.; Sasidhara, S. *Hot Working Guide: A Compendium of Processing Maps*, 2nd ed.; ASM International: Materials Park, OH, USA, 2015; ISBN 978-1-62708-091-0.
21. Prasad, Y.V.R.K.; Seshacharyulu, T. Modeling of hot deformation for microstructural control. *Inter. Mater. Rev.* **1998**, *44*, 243–258. [CrossRef]
22. Prasad, Y.V.R.K. Processing Maps—A Status Report. *J. Mater. Eng. Perform.* **2003**, *12*, 638–645. [CrossRef]
23. Ziegler, H. Some extremum principles in irreversible thermodynamics with applications to continuum mechanics. In *Progress in Solid Mechanics*; John Wiley: New York, NY, USA, 1965; Volume 4, pp. 91–193.
24. Jonas, J.J.; Sellars, C.M.; Tegart, W.M. Strength and structure under hot working conditions. *Met. Rev.* **1969**, *14*, 1–24. [CrossRef]
25. Prasad, Y.V.R.K.; Rao, K.P. Processing maps and rate controlling mechanisms of hot deformation of electrolytic tough pitch copper in the temperature range 300–950 °C. *Mater. Sci. Eng. A* **2005**, *391*, 141–150. [CrossRef]
26. Böttger, B.; Eiken, J.; Ohno, M.; Klaus, G.; Fehlbier, M.; Schmid-Fetzer, R.; Steinbach, I.; Bührig-Polacz, A. Controlling microstructure in magnesium alloys: A combined thermodynamic, experimental and simulation approach. *Adv. Eng. Mat.* **2006**, *8*, 241–247. [CrossRef]
27. Grobner, J.; Schmid-Fetzer, R. Key issues in a thermodynamic Mg alloy database. *Metall. Mater. Trans. A* **2013**, *44*, 2918–2934. [CrossRef]
28. Laser, T.; Nurnberg, M.R.; Janz, A.; Hartig, C.; Letzig, D.; Schmid-Fetzer, R.; Bormann, R. The influence of manganese on the microstructure and mechanical properties of AZ31 gravity die cast alloys. *Acta Mater.* **2006**, *54*, 3033–3041. [CrossRef]
29. Frost, H.J.; Ashby, M.F. *Deformation-Mechanism Maps: The Plasticity and Creep of Metals and Ceramics*; Pergamon Press: Oxford, UK, 1982; p. 44.
30. Morris, J.R.; Schraff, J.; Ho, K.M.; Turner, D.E.; Ye, Y.Y.; Yoo, M.H. Prediction of a {11-22} hcp stacking fault using a modified generalized stacking-fault calculation. *Philos. Mag. A* **1997**, *76*, 1065–1077. [CrossRef]
31. Hilborn, R.C. *Chaos and Non-Linear Dynamics*; Oxford University Press: New York, NY, USA; Oxford, UK, 1994; p. 625.

Article

Relationship among Initial Texture, Deformation Mechanism, Mechanical Properties, and Texture Evolution during Uniaxial Compression of AZ31 Magnesium Alloy

Hui Su [1], Zhibing Chu [1,2,*], Chun Xue [1], Yugui Li [1] and Lifeng Ma [1]

1 Engineering Research Center Heavy Machinery Ministry of Education, Taiyuan University of Science and Technology, Taiyuan 030024, China; suhui941229@163.com (H.S.); xuechun999@126.com (C.X.); liyugui2008@163.com (Y.L.); mlf_tyust@163.com (L.M.)

2 School of Mechanics and Architectural Engineering, Jinan University, Guangzhou 510632, China

* Correspondence: chuzhibing@tyust.edu.cn; Tel.: +86-184-3531-8786

Received: 27 July 2020; Accepted: 18 August 2020; Published: 21 August 2020

Abstract: Cuboid samples with significant initial texture differences were cut from extruded AZ31 Mg alloy samples, whose long axis and bar extrusion direction ED were 0° (sample E0), 45° (sample E45), and 90° (sample E90). The relationship among the initial texture, deformation mechanism, mechanical properties, and texture evolution of the AZ31 Mg alloy was investigated systematically using a compression test, microstructure characterization, and the Viscoplastic Self-Consistent (VPSC) model. Results revealed a close relationship among them. By influencing the activation of the deformation mechanism, the deformation under different initial textures resulted in obvious mechanical anisotropy. Compared with E0 and E90, the initial texture of E45 was more conducive to the improvement of reforming ability after pre-compression. Meanwhile, the initial texture significantly affected the microstructure characteristics of the material, especially the number and morphology of the {10–12} tensile twins. Texture results showed that the priority of deformation mechanism depended on the initial texture and led to the difference in texture evolution.

Keywords: initial texture; deformation mechanism; mechanical properties; texture evolution

1. Introduction

Mg alloy has high specific stiffness, high specific strength, low density, and good damping performance [1], hence an important choice for material application in the fields of communication, military industry, and automobile [2–4]. AZ31 is one of the most widely applied variants among Mg alloys. Experimental studies showed that the initial texture in the forming process of Mg alloy has a remarkable effect on its material properties [5–10].

The rolling direction (RD) and transverse direction (TD) samples with distinct differences in initial texture were subsequently subjected to the bending process that has been reported by Wang [11] et al., and the results revealed that the conspicuous asymmetry in the bending behavior is ascribed to the initial texture and weaker basal texture is favorable to improving the bending properties. Singh [12] investigated and compared the micromechanical deformation behavior of E-form fine grain (EFG), E-form coarse grain (ECG), and AZ31 Mg alloys by using a mini V-bending test. The EFG and ECG Mg alloys with weaker texture have better bendability than the AZ31 alloys with stronger texture. Xiong [13] investigated the asymmetrical planar mechanical behavior of an extruded AZ31 sheet possessing bimodal texture distribution. The results indicated that the plastic deformation mechanism activated in the c-axis//TD textured grains during rolling is sensitive to the rolling paths. Therefore, initial texture plays an important role in the plastic deformation of Mg alloys. Wang [11] pointed out that the selection of a dominant mechanism is highly dependent on the initial texture of Mg alloys, including the activation of {10–12} tension twins upon compression perpendicular or tension parallel to the c-axis. Consequently, the difference in initial texture affects the twin deformation [14–17], and the deflection of crystal orientation caused by twinning makes the texture evolution more complicated [18–21]. Therefore, the relationship among initial texture, deformation mechanism, mechanical properties, and texture evolution during plastic deformation of Mg alloy must be studied.

The most commonly used technical methods for analyzing and studying material texture at present are X-ray diffraction (XRD) and electron backscattering diffraction (EBSD) [22]. However, the shortcomings of these two experimental methods are their high cost of testing, the difficulty of sample preparation, and the long experimental period. On the basis of the common optical reflection method, Gaskey [23] realized the grain orientation characterization of metal Ni and semiconductor Si wafers under the directional reflection microscope (DRM). However, the disadvantage of this experimental method is its inability to quantitatively study the effects of slip and twinning on texture evolution. The proposed Viscoplastic Self-Consistent (VPSC) model [24,25] solves the problem of studying metal plastic deformation from the perspective of microscopic deformation mechanism. Su [26] et al. used the VPSC model with different deformation mechanisms to investigate the influence of a secondary deformation mechanism (prismatic <a> slip, pyramidal <c+a> slip, and {10 $\bar{1}$1} compression twin) on mechanical response and texture evolution, and a VPSC model was established to simulate the plastic deformation of magnesium alloy.

Chapuis [27] used VPSC to simulate the plastic deformation and deformed texture of Mg-3Al-1Zn plate and compared them with the experimental observations. The results demonstrated that the {10–12} extension twins are responsible for the main texture variations. Zhang [28] used VPSC to investigate the in-plane anisotropy of a Mg alloy AZ31B-O sheet. The hexagonal close-packed crystallographic structure, deformation twinning, and initial basal texture are responsible for the characteristic behavior of Mg alloys. Alireza [29] also used VPSC to investigate the effect of alloying elements on Mg's tensile behavior, specifically the relative activity of different slip and twinning modes. The VPSC model has been successfully applied to predict the mechanical behavior and deformation texture of Mg alloys under different plastic deformations [30].

In this paper, E0, E45, and E90 cuboid samples along the long axis and extrusion direction of 0°, 45°, and 90° were cut from the extruded AZ31 Mg alloy bar for a uniaxial compression experiment at room temperature. The VPSC model and Predominant Twin Reorientation (PTR) scheme [31] were used to simulate the mechanical behavior and texture evolution of the different initial texture samples under uniaxial compression, and the anisotropic behavior and texture evolution differences in Mg alloy samples were analyzed from the perspective of a microscopic deformation mechanism. The microstructure and deformed texture were characterized by EBSD and compared with the simulation results. The relationship among initial texture, deformation mechanism, mechanical properties, and texture evolution were studied.

2. Materials and Methods

2.1. Sample Preparation

The extrusion direction of magnesium alloy is expressed by ED, and the transverse and normal directions perpendicular to the extrusion direction are expressed by TD and ND, respectively. In addition, 8 mm × 8 mm × 12 mm cuboids with different orientations were cut from extruded AZ31 Mg alloy samples, as shown in Figure 1b, whose long axis and bar extrusion ED were 0° (sample E0), 45° (sample E45), and 90° (sample E90), as shown in Figure 1a. Compression tests at a strain rate of 10^{-3} s^{-1} and loading direction along the long axis were carried out on the electronic universal testing machine (model: WDW-E100D) provided by Changchun new testing machine Co., Ltd in Jilin Province of China at room temperature.

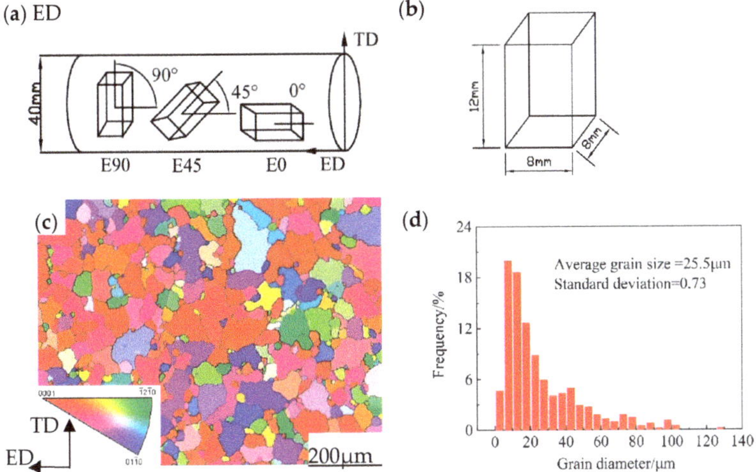

Figure 1. Sample shape and size: (**a**) schematic of samples; (**b**) compression samples; (**c**) inverse pole figure map; (**d**) grain size distribution.

The original and compressed samples were cut to 10 × 7 × 6 mm in length, width, and height to avoid damaging the electron microscope lens. After plastic deformation, EBSD was conducted on the JEOL JSM-7800F field emission scanning electron microscope provided by Oxford Instrument Technology (Shanghai) Co., Ltd in China to analyze the microscopic structure, crystal orientation, and structural characteristics of the sample. Mechanical grinding and electrochemical polishing were finished when preparing EBSD. In mechanical grinding, the sample was polished to 2000 # with sandpaper until there was no scratch on the surface. Then, electrolytic polishing was carried out. AC2 solution is used as a polishing solution, and the formula is shown in Table 1. The polishing parameters are as follows: voltage 20 V, temperature −30 °C, polishing time 120 s, polishing current 0.03–0.08 A. The sample was cleaned with alcohol after polishing. The voltage of the EBSD data collection is 20 kV, the inclination angle is 70°, the current is 15 mA, and the working distance is 15 mm. Channel 5 software is used to process the data collected.

Table 1. The prescription of AC2.

AC2 I	800 mL ethanol, 100 mL propanol, 18.5 mL distilled water, 10 g hydroxyquimoline, 75 g citric acid
AC2 II	41.5 g sodium thiocyanate
AC2 III	15 mL perchloric acid

The microstructure and grain size of the longitudinal section (ED–TD plane) of E0 sample were characterized by EBSD, as shown in Figure 1c. Most of the initial samples had an equiaxed crystal structure, and the average grain size was 25.5 μm, as shown in Figure 1d. According to the calculation formula of standard deviation (SD), the SD was 0.73. Besides, some elongated grains were revealed due to the grain stretching along the ED during extrusion.

2.2. VPSC-PTR Model

The VPSC model considers the interaction between grains in polycrystals and assumes that the grains are ellipsoid, and the model adopts the following rate-dependent continuous constitutive model equation [24]:

$$\varepsilon_{ij}(\overline{x}) = \sum_s m_{ij}^s \gamma^s(\overline{x}) = \gamma_0 \sum_s m_{ij}^s \left\{ \frac{m_{kl}^s \sigma_{kl}(\overline{x})}{\tau^s} \right\}^n = M_{ijkl}\sigma_{kl}(\overline{x}) \tag{1}$$

where τ^s denotes the critical shear stress, $m_{ij}^s = \frac{1}{2}(n_i^s b_j^s + b_j^s n_i^s)$ is the Schmid factor of the slip system/twin system(s), n^s and b^s represent the normal direction (ND) of the slip/twin plane and the slip/twin direction, respectively. $\varepsilon_{ij}(\overline{x})$ and $\sigma_{kl}(\overline{x})$ are the strain and stress partial tensors, respectively. γ^s stands for the local shear rate acting on the slip system(s). γ_0 represents the normalized coefficient, and n is the rate sensitivity index. M_{ijkl} stands for the viscoplastic convention and can relate the macroscopic strain rate to the macroscopic deviatoric stress.

The critical shear stress τ^s is formed with the accumulation of shear amount in each grain. Therefore, the evolution rules can be described by the Voce hardening model:

$$\tau^s = \tau_0^s + (\tau_1^s + \theta_1^s \Gamma)(1 - \exp(-\frac{\theta_0^s \Gamma}{\tau_1^s})) \tag{2}$$

where $\Gamma = \sum_s \Delta\gamma^s$ is the cumulative shear of the grain, τ_0 denotes the initial critical stress, θ_0 stands for the initial hardening rate, θ_1 refers to the saturation hardening rate, τ_0^s and τ_1^s are the initial and extrapolated critical shear stresses of mechanisms, respectively. Figure 2 depicts the specific definitions.

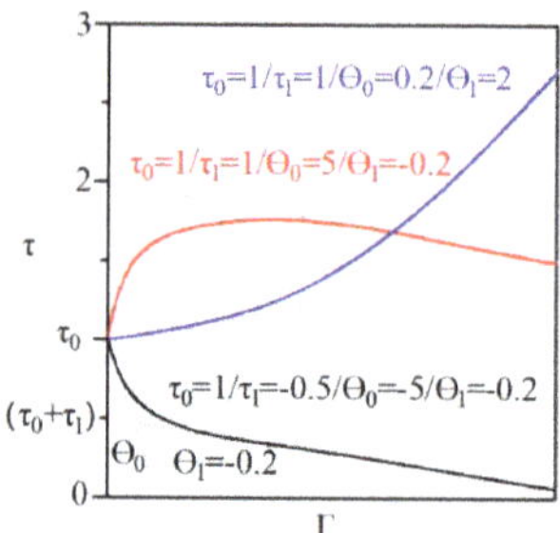

Figure 2. Physical meaning of hardening parameters for Viscoplastic Self-Consistent (VPSC) simulations.

In this study, PTR [31] is used to analyze the effect of twins on texture evolution during plastic deformation. For grain g, $\gamma^{t,g}$ is the shear strain caused by each twin t, and the corresponding twin volume fraction is $V^{t,g} = \gamma^{t,g}/S^t$, where S^t is the intrinsic shear strain of twin system t.

The integral number of twins caused by all twin systems in all grains is called the cumulative twin integral number:

$$V^{acc,mode} = \sum_g \sum_t \gamma^{t,g}/S^t \tag{3}$$

Through superimposition of the twin fraction of each incremental step, the cumulative twin fraction is compared with the critical volume fraction caused by twins. Thus, the critical volume fraction is defined as:

$$V^{th,mode} = A^{th1} + A^{th2}\frac{V^{eff,mode}}{V^{acc,mode}} \tag{4}$$

where $V^{eff,mode}$ is the effective twin crystal integral number, A^{th1} and A^{th2} are material constants.

3. Results and Discussion

The input data on the crystallographic orientation required for VPSC modeling were extracted from the EBSD map measured in Figure 1 through the orientation distribution function used by the toolbox MTEX in MATLAB code, as described in the literature [32]. A total of 2000 discrete directions with the same volume fraction were used for VPSC modeling. The input data required for modeling were described in detail in the literature [26], hence not discussed here. The detailed procedure for the calculation in the VPSC model has been well documented by Li et al. [33].

Given the different dominant deformation mechanisms of Mg alloys under different loading modes, according to Figure 2, the hardening parameters of deformation mechanisms could be obtained by simulating the stress-strain curves of specific loading experiments. The specific fitting method is as follows: first, the hardening parameters of basal slip and prismatic slip were determined by fitting the compression results of E90. Then, the hardening parameters of pyramidal slip were determined by fitting the compression results of E45. Finally, the hardening parameters of tensile and compression twinnings were determined using the compression results of E0. The results of the fitting curve are shown in Figure 3. The differences in initial texture resulted in obvious mechanical anisotropy in the stress-strain curves of various samples. In the initial pole figure of E0, most of the c-axis of the grain was perpendicular to the ED; that is, the basal was parallel to the ED, resulting in a strong basal texture in the initial pole figure distribution of Mg alloy, as shown in Figure 3c. When the E45 sample was compressed, the initial texture was rotated along the ED direction by 45°, and the resulting texture is shown in Figure 3b. Similarly, the macroscopic strain rate and stress along the TD directions were the loading conditions of the E90 sample, as shown in Figure 3a. Analysis of the stress-strain curves of the different samples indicated that the stress-strain curves of E45 and E90 were similar to "S," and the flow stress of E90 was higher than that of E45. The stress-strain curve of E0 had a certain s-shaped characteristic, but it was not significant, and its yield strength was slightly higher than that of E45 and E90. The stress-strain curves predicted by the VPSC model were in good agreement with the experimental results and accurately reflected the stress-strain characteristics of each stage in the axial tension and compression plastic deformation process.

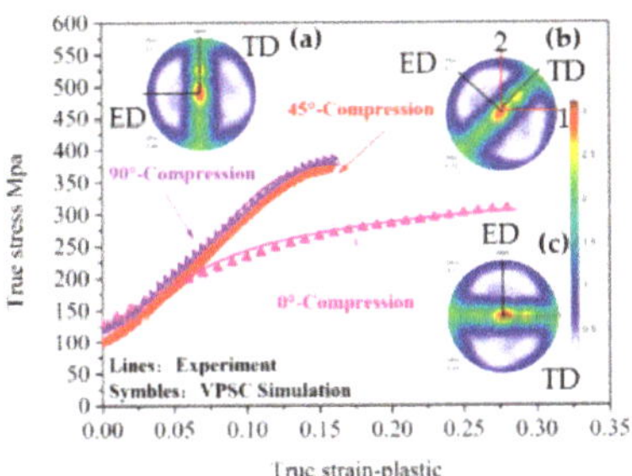

Figure 3. Experimental (symbol) and simulated (line) stress-strain curves under uniaxial compression and corresponding initial texture of (**a**) E0; (**b**) E45; (**c**) E90.

The optimal hardening parameters of different deformation mechanisms were determined, as shown in Table 2. Results show that at room temperature, the critical resolved shear stress (CRSS) of basal <a> slip [18] (approximately 0.45–0.81 MPa) < the CRSS of the {10-12} tension twin [19] (approximately 2–2.8 MPa) < the CRSS of the prismatic <a> slip [34] (approximately 39.2 MPa) < the CRSS of the pyramidal <c+a> slip [35] (approximately 45–81 MPa) < the CRSS of the {10-11} compression twin [25] (approximately 76–153 MPa) of magnesium and its alloys. Table 2 presents that the CRSS of the overall size order is consistent with the research value. Optimal hardening parameters play an important guiding role for future researchers to study the macro- and micro-deformation mechanisms of magnesium alloys.

Table 2. Hardening parameters of each slip and twinning system in the simulation of Mg alloy under monotonic loading.

Deformation Mode	τ_0/MPa	τ_1/MPa	θ_0/MPa	θ_1/MPa
Basal <a>	28	60	185	16
Prismatic <a>	85	10	200	250
Pyramidal <c+a>	110	15	120	400
Extension twin	45	0	150	400
Compression twin	210	100	345	400

Figure 4 shows the prediction results of the relative activating amount of the deformation mechanism during the compression deformation of samples with different initial textures. At the initial stage of deformation, the deformation mechanism of E0 was dominated by tensile twin, supplemented by basal slip. The dominant deformation mode of E45 and E90 was basal slip, supplemented by tensile twin, while CRSS tensile twin > CRSS basal slip. As a result, the yield stress of E0 was slightly greater than that of E45 and E90. Figure 4a illustrates that when E0 was compressed, the activation of the tensile twin decreased as the compression amount continued to increase, and the activation of the basal, prismatic, and pyramidal slips increased. This finding was attributed to the twin growth and fusion behavior of the specimen in the later stage of deformation [19], resulting in a surface phenomenon, wherein the twin volume fraction decreases due to the disappearance of twin boundaries. Figure 4b,c show that E45 and E90 also had a small role in the activation of tensile twins at the initial deformation stage. During the compression process of E45 and E90 samples, the c-axis of the grain was in a state of tensile stress for the part of the grain nearly perpendicular to the direction of the application, which was conducive to the activation of tensile twins. With the increase in strain, the activity of the pyramidal slip of E90 was significantly greater than that of E45, while the pyramidal slip of CRSS was higher. This difference was reflected in the stress-strain curve, where the flow stress of E90 was higher than that of E45 [22].

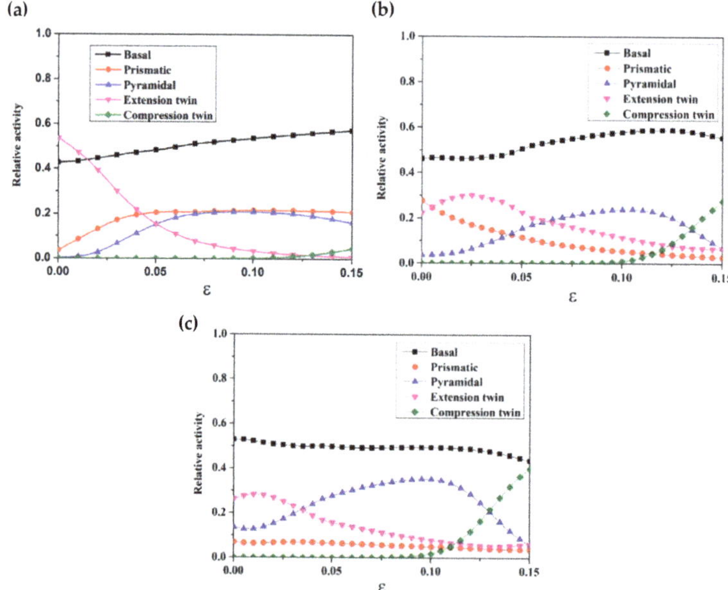

Figure 4. Predicted relative activity of deformation modes under uniaxial compression of (**a**) E0; (**b**) E45; (**c**) E90.

Figure 5a–c shows the predicted and experimental texture with axial compression of E0, E45 and E90 samples to 8% by using the VPSC model. Through comparison, it can be seen that the simulation results with 2000 gains are in good agreement with the experimental results, which reasonably reflects the texture evolution during compression of the E0, E45 and E90 samples and verified the accuracy of the VPSC model in simulating the texture evolution of magnesium alloy in this study.

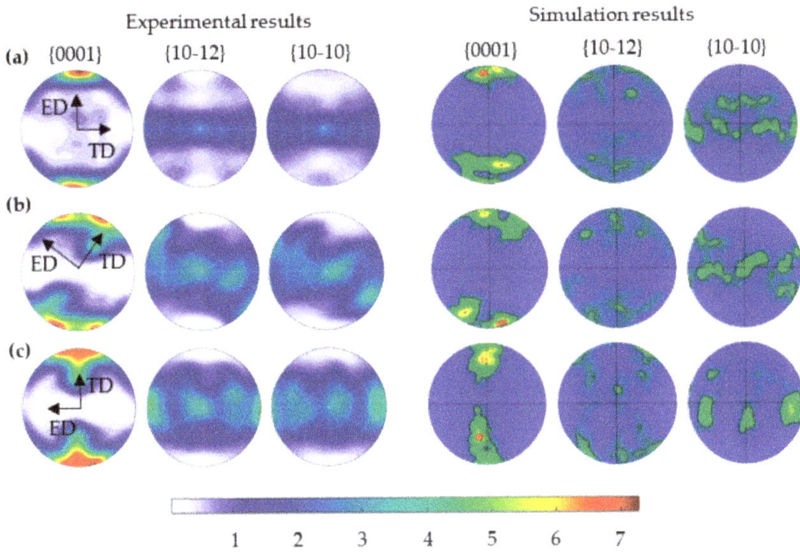

Figure 5. Experimental and Simulation pole figures under uniaxial compression to 8% of (**a**) E0 sample; (**b**) E45 sample; (**c**) E90 sample.

Figure 6 shows the prediction results of {0001} pole figure during uniaxial compression of the samples with different initial textures. As shown in Figure 6a, when E0 was compressed to $\varepsilon = 0.02$, a large peak value was rapidly formed in the TD direction of the {0001} grain, while the peak value near the ND direction was weakened due to the activation of {10–12} tensile twins (Figure 3a). This activation resulted in a mutation of ~90° in the grain c-axis, with {0001} basal deflected from parallel to compression direction to perpendicular to compression direction. With the increase in strain, the number of tensile twins increased, the peak value in the TD direction continued to increase, and the strength peak near the ND direction gradually disappeared. Besides, the orientation peak dispersion in the TD direction of {0001} surface increased slightly with the increase in strain, possibly due to the formation of new grains at the intersection of the twins brought by dynamic recrystallization similar to that in the original tensile twins but with slight deflection [36,37]. As shown in Figure 6b, when E45 was compressed to $\varepsilon = 0.04$, the peak value of the strength of {0001} plane turned to near the direction of force application (direction 2). The reason was because when it was compressed along direction 2, the deformation mechanism at the initial stage of deformation was dominated by basal slip (Figure 4b), which could not change the grain orientation, and the grain rotated slowly under the restrictions of loading direction, structural integrity, and deformation continuity. With the increase in stress, the shear stress of the tensile twin started to activate upon reaching the corresponding shear stress, making the texture gradually change from the original position of 45° from the compression direction to the ED direction. Figure 6c shows that the dominant deformation mechanism of E90 at the initial deformation stage was the basal slip (Figure 4c), most of the grains rotated continuously and gradually, and the normal of {0001} basal rotated parallel to the compression axis. With the increase in strain, the peak strength was always near the TD direction. The disappearance of the polar density in the center of the map was the result of the mutation of grain orientation caused by the effect of tensile twin.

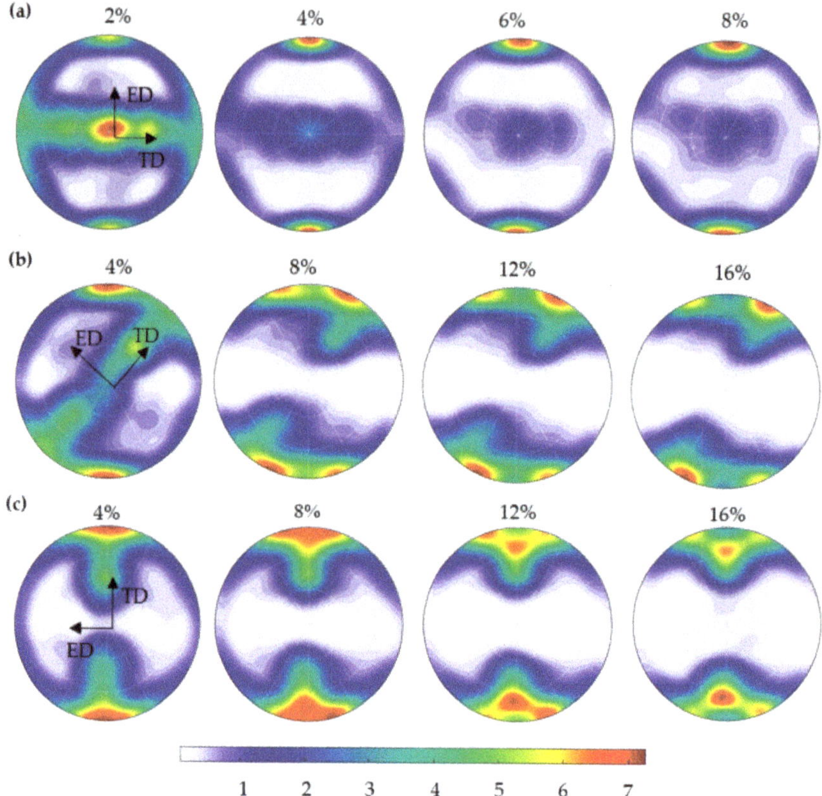

Figure 6. Predicted texture evolution of {0001} pole figure under uniaxial compression to different strain levels of (**a**) E0; (**b**) E45; (**c**) E90.

In summary, for E0 and E90 after uniaxial compression plastic deformation, the {0001} surface of most grains either turned to the compression direction or remained perpendicular to the compression direction all the time. The {0001} surface of this part of the grain was always in hard orientation, which was not conducive to the activation of slip. Thus, the reforming ability after pre-compression could not be improved. In the uniaxial compression of the E45 sample, the initial deformation was dominated by basal slip, and the grain c-axis rotated slowly parallel to the direction of the application of force, which was beneficial for improving the reforming ability after pre-compression.

Figure 7a–c shows the grain boundary structure maps, typical grain EBSD maps, {0001} pole figure, and grain boundary misorientation maps of the different initial texture samples when they were uniaxially compressed to 0.08. In the grain boundary structure maps, the grain boundary satisfying the orientation relation of 86.3° ± 5° was defined as the stretched twin grain boundary, which was represented by the solid red line. Meanwhile, low-angle grain boundaries (LAGBs) and high-angle grain boundary (HAGBs) fractions could be confirmed by observing the misorientation boundary fraction from EBSD. In E0, the color of the EBSD maps of grain Ma showed that grain Ma underwent twin growth and fusion behavior. This finding also explained the illusion that the tensile twins decrease or even disappear in the later stage of compression deformation along the ED direction [38]. The pole figure of E45 and E90 showed that the basal plane of parent grain Mb and Mc tended to be parallel to ED direction and was located near the center of the polar diagram. In addition, the activation of tensile twins caused the grain c-axis to deflect ~90°, as shown in the twin Tb and Tc in the pole figure. Comparison of the experimental and simulation results of the different sample pole figures demonstrated that a small amount of polar density still remained in the center of the experimental pole figure of E45 and E90. The reason was because in the VPSC model, the PTR scheme assumes that when the cumulative twin fraction exceeds the critical volume fraction, the entire grain orientation changes, thereby ignoring the parent grain orientation. The disappearance of polar density in the center of E0 again proved the growth and fusion behavior of the stretched twins. When comparing the grain boundary misorientation maps of different samples, a slight decrease in LAGBs may be ascribed to the increased fraction of the {10–12} tension twin in E90. This finding indicated that the main deformation mechanism in E90 was the tension twin instead of the slip, whose activation could produce profuse LAGBs.

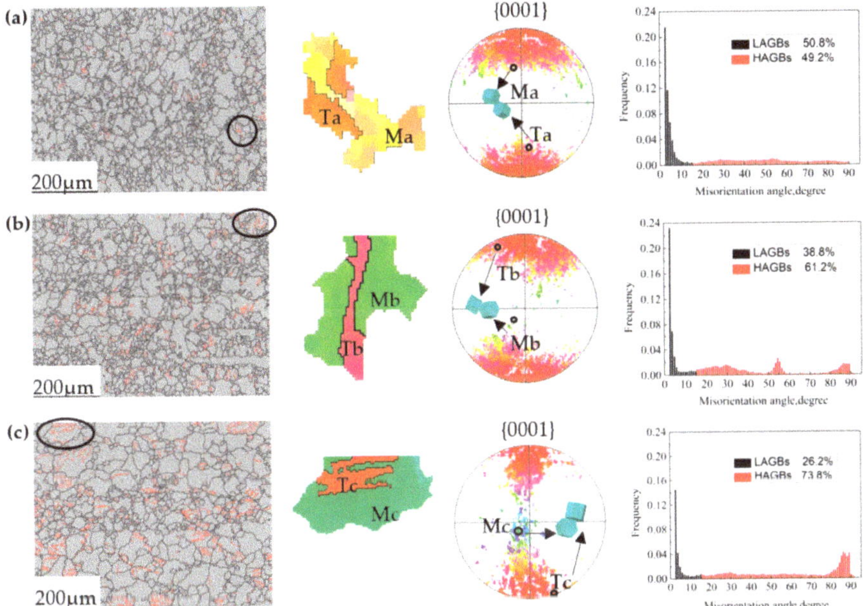

Figure 7. Boundary structure maps, electron backscattering diffraction (EBSD) maps of typical grain, {0001} pole figure, and grain boundary misorientation map under uniaxial compression to 8% of (**a**) E0; (**b**) E45; (**c**) E90. (LAGBs: low angle grain boundaries between 0° and 15°; HAGBs: high angle grain boundaries higher than 15°).

4. Conclusions

On the basis of the VPSC model and experimental methods, the relationship among initial texture, deformation mechanism, mechanical behavior, and texture evolution during the plastic deformation of Mg alloy was studied, and the following conclusions were drawn:

(1) The obvious difference in initial texture led to obvious anisotropy of the compression behavior due to the influence in the activation of the deformation mechanism. At the initial stage of deformation, the compressive yield strength of E0, with tensile twins as the dominant deformation mechanism, was slightly higher than that of E45 and E90. The activity of the pyramidal slip was high in the compression process of E90, thereby leading to high flow stress.

(2) Initial texture played an important role in texture evolution during compression by influencing the priority of deformation mechanism. The large activation of the tensile twin in E0 made the {0001} basal plane deflects perpendicular to the compression direction. The activation of the basal slip of E45 and E90 caused the grain c-axis to rotate slowly parallel to the direction of application, and the initiation of the tensile twin caused the {0001} plane to slowly turn perpendicular to the ED direction. Compared with E0 and E90, E45 was more conducive to the improvement of reforming ability after pre-compression.

(3) The difference in microstructure among the three samples was attributed to the number and morphology of the {10–12} tensile twins, which were mainly due to the obvious difference in initial texture.

Author Contributions: Writing—original draft preparation, H.S.; writing—review and editing, Z.C.; Helping the first author to build the VPSC model, C.X.; project administration, Y.L.; Helping the first author to conduct the experiment, L.M. All authors have read and agreed to the published version of the manuscript.

Funding: This research was funded by National Key Research and Development Program of China (No. 2018YFB1307902), National Science Fund Subsidized Project (No. U1710113), Shanxi Province Joint Student Training Base Talent Training Project (No. 2018JD33), Shanxi Excellent Youth Fund (No. 201901d211312), Transformation and cultivation project of scientific and technological achievements in Colleges and universities of Shanxi Province (No. 2019KJ028), Shanxi Graduate Education Innovation Project (No. 2019SY482).

Acknowledgments: The authors express their thanks to the people helping with this work, and acknowledge the valuable suggestions from the peer reviewers.

Conflicts of Interest: The authors declare no conflict of interest.

References

1. Wan, D. Near spherical α-Mg dendrite morphology and high damping of low-temperature casting Mg–1wt.%Ca alloy. *Mater. Charact.* **2011**, *62*, 8–11.
2. Pan, H.; Qin, G.; Huang, Y.; Ren, Y.; Sha, X.; Han, X.; Liu, Z.; Li, C.; Wu, X.; Chen, H.; et al. Development of low-alloyed and rare-earth-free magnesium alloys having ultra-high strength. *Acta Mater.* **2018**, *149*, 350–363. [CrossRef]
3. Zeng, Z.; Stanford, N.; Davies, C.H.J.; Nie, J.-F.; Birbilis, N. Magnesium extrusion alloys: A review of developments and prospects. *Int. Mater. Rev.* **2018**, 1–36. [CrossRef]
4. Lu, X.; Zhao, G.Q.; Zhou, J.X.; Zhang, G.S.; Sun, L. Effect of extrusion speeds on the microstructure, texture and mechanical properties of high-speed extrudable Mg-ZnSn-Mn-Ca alloy. *Vacuum* **2018**, *157*, 180–191. [CrossRef]
5. Wang, H.; Lu, C.; Tieu, A.K.; Deng, G. Coupled effects of initial orientation scatter and grain-interaction to texture evolution: A crystal plasticity FE study. *Int. J. Mater. Form.* **2019**, *12*, 161–171. [CrossRef]
6. Rodrigues, D.G.; de Alcântara, M.C.; de Oliveira, T.R.; Gonzalez, B.M. The effect of grain size and initial texture on microstructure, texture, and formability of Nb stabilized ferritic stainless steel manufactured by two-step cold rolling. *J. Mater. Res. Technol.* **2019**, *8*, 4151–4162. [CrossRef]
7. Wang, X.; Shi, T.; Jiang, Z.; Chen, W.; Guo, M.; Zhang, J.; Zhuang, L.; Wang, Y. Relationship among grain size, texture and mechanical properties of aluminums with different particle distributions. *Mater. Sci. Eng.* **2019**, *753*, 122–134. [CrossRef]
8. He, J.; Mao, Y.; Gao, Y.; Xiong, K.; Jiang, B.; Pan, F. Effect of rolling paths and pass reductions on the microstructure and texture evolutions of AZ31 sheet with an initial asymmetrical texture distribution. *J. Alloy. Compd.* **2019**, *786*, 394–408. [CrossRef]
9. Singh, J.; Kim, M.S.; Choi, S.H. The effect of initial texture on deformation behaviors of Mg alloys under erichsen test. *Magnes. Technol.* **2018**, 223–229. [CrossRef]
10. Ahmadian, P.; Abbasi, S.M.; Morakabati, M. Effect of initial texture and grain size on geometrically necessary dislocations density distribution during uniaxial compression of Ti-6Al-4V. *Materialstoday* **2018**, *14*, 263–272. [CrossRef]

11. Wang, W.; Zhang, W.; Chen, W.; Cui, G.; Wang, E. Effect of initial texture on the bending behavior, microstructure and texture evolution of ZK60 magnesium alloy during the bending process. *J. Alloy. Compd.* **2018**, *737*, 505–514. [CrossRef]
12. Singh, J.; Kim, M.S.; Choi, S.H. The effect of initial texture on micromechanical deformation behaviors in Mg alloys under a mini-V-bending test. *Int. J. Plast.* **2019**, *117*, 33–57. [CrossRef]
13. Xiong, Y.; Gong, X.; Jiang, Y. Effect of initial texture on fatigue properties of extruded ZK60 magnesium alloy. *Fatigue Fract. Eng. Mater. Struct.* **2018**, *41*, 1504–1513. [CrossRef]
14. Wang, H.; Wu, P.D.; Wang, J.; Tome, C.N. A physics-based crystal plasticity model for hexagonal close packed (HCP) crystals including both twinning and de-twinning mechanisms. *Int. J. Plast.* **2013**, *49*, 36–52. [CrossRef]
15. Wu, P.D.; Guo, X.Q.; Qiao, H.; Agnew, S.R.; Lloyd, D.J.; Embury, J.D. On the rapid hardening and exhaustion of twinning in magnesium alloy. *Acta Mater.* **2017**, *122*, 369–377. [CrossRef]
16. Balik, J.; Dobron, P.; Chmelik, F.; Kuzel, R.; Drozdenko, D.; Bohlen, J.; Letzig, D.; Lukac, P. Modeling of thework hardening in magnesium alloy sheets. *Int. J. Plast.* **2015**, *76*, 166–185. [CrossRef]
17. Barnett, M.R.; Ghaderi, A.; da Fonseca, J.Q.; Robson, J.D. Influence of orientation on twin nucleation and growth at low strains in a magnesium alloy. *Acta Mater.* **2014**, *80*, 380–391. [CrossRef]
18. Lou, X.Y.; Li, M.; Boger, R.K.; Agnew, S.R.; Wagoner, R.H. Hardening evolution of AZ31B Mg sheet. *Int. J. Plast.* **2007**, *23*, 44–86. [CrossRef]
19. Koike, J. Enhanced deformation mechanisms by anisotropic plasticity in polycrystalline Mg alloys at room temperature. *Metall. Mater. Trans. A* **2005**, *36*, 1689–1696. [CrossRef]
20. Lou, C.; Zhang, X.; Ren, Y. Non-Schmid-based 10–12 twinning behavior in polycrystalline magnesium alloy. *Mater. Charact.* **2015**, *107*, 249–254. [CrossRef]
21. Huang, H.; Godfrey, A.; Zheng, J.; Liu, W. Influence of local strain on twinning behavior during compression of AZ31 magnesium alloy. *Mater. Sci. Eng. A* **2015**, *640*, 330–337. [CrossRef]
22. Nicolas, B.; Hendrix, D.; Raynald, G. Imaging with a Commercial Electron Backscatter Diffraction (EBSD) camera in a scanning electron microscope: A review. *J. Imaging* **2018**, *4*, 88.
23. Gaskey, B.; Hendl, L.; Xiaogang, W.; Seita, M. Full length article Optical characterization of grain orientation in crystalline materials. *Acta Mater.* **2020**, *194*, 558–564. [CrossRef]
24. Wang, H.; Wu, P.D.; Tomé, C.N.; Huang, Y. A finite strain elastic-viscoplastic self-consistent model for polycrystalline materials. *J. Mech. Phys. Solids* **2010**, *58*, 594–612. [CrossRef]
25. Lebensohn, R.; Gonzalez, M.; Tomé, C.; Pochettino, A. Measurement and prediction of texture development during a rolling sequence of zircaloy-4 tubes. *J. Nucl. Mater.* **1996**, *229*, 57–64. [CrossRef]
26. Su, H.; Chu, Z.B.; Wang, H.Z.; Li, Y.C.; Ma, L.F.; Xue, C. Mechanism of secondary deformation of extruded AZ31 magnesium alloy by viscoplastic self-consistent model. *Adv. Mater. Sci. Eng.* **2020**, 1–11. [CrossRef]
27. Zhao, L.; Chapuis, A.; Xin, Y.; Liu, Q. VPSC-TDT modeling and texture characterization of the deformation of a Mg-3Al-1Zn plate. *J. Alloy. Compd. Interdiscip. J. Mater. Sci. Solid State Chem. Phys.* **2017**, *710*, 159–165. [CrossRef]
28. Zhang, B.; Li, S.; Wang, H.; Tang, W.; Jiang, Y.; Wu, P. Investigation of the in-plane mechanical anisotropy of magnesium alloy AZ31B-O by VPSC–TDT crystal plasticity model. *Materials* **2019**, *12*, 1590. [CrossRef]
29. Maldar, A.; Wang, L.; Zhu, G.; Zeng, X. Investigation of the alloying effect on deformation behavior in Mg by Visco-Plastic Self-Consistent modeling. *J. Magnes. Alloy.* **2020**, *14*, 13–42. [CrossRef]
30. Chaudry, U.M.; Hamad, K.; Kim, J.-G. Ca-induced plasticity in magnesium alloy: EBSD measurements and VPSC calculations. *Crystals* **2020**, *10*, 67. [CrossRef]
31. Agnew, S.R.; Brown, D.W.; Tomé, C.N. Validating a polycrystal model for the elastoplastic response of magnesium alloy AZ31 using in situ neutron diffraction. *Acta Mater.* **2006**, *54*, 4841–4852. [CrossRef]
32. Hu, L.; Jiang, S.; Zhou, T.; Chen, Q. A coupled finite element and crystal plasticity study of friction effect on texture evolution in uniaxial compression of niti shape memory alloy. *Materials* **2018**, *11*, 2162. [CrossRef]
33. Li, H.; Zhang, H.; Yang, H.; Fu, M.; Yang, H. Anisotropic and asymmetrical yielding and its evolution in plastic deformation: Titanium tubular materials. *Int. J. Plast.* **2017**, *90*, 177–211. [CrossRef]
34. Reed-Hill, R.E.; Robertson, W.D. Additional modes of deformation twinning in magnesium. *Acta Metall.* **1957**, *5*, 717–727. [CrossRef]
35. Agnew, S.R.; Yoo, M.H.; Tomé, C.N. Application of texture simulation to understanding mechanical behavior of Mg and solid solution alloys containing Li or Y. *Acta Mater.* **2001**, *49*, 4277–4289. [CrossRef]

36. Al-Samman, T.; Gottstein, G. Dynamic recrystallization during high temperature deformation of magnesium. *Mater. Sci. Eng. A* **2008**, *490*, 411–420. [CrossRef]
37. Li, X.; Yang, P.; Wang, L.-N.; Meng, L.; Cui, F. Orientational analysis of static recrystallization at compression twins in a magnesium alloy AZ31. *Mater. Sci. Eng. A* **2009**, *517*, 160–169. [CrossRef]
38. Wu, W.X.; Jin, L.; Dong, J.; Ding, W.J. Deformation behavior and texture evolution in an extruded Mg-lGd alloy during uniaxial compression. *Mater. Sci. Eng. A* **2014**, *593*, 48–54. [CrossRef]

MDPI
St. Alban-Anlage 66
4052 Basel
Switzerland
Tel. +41 61 683 77 34
Fax +41 61 302 89 18
www.mdpi.com

Crystals Editorial Office
E-mail: crystals@mdpi.com
www.mdpi.com/journal/crystals

www.ingramcontent.com/pod-product-compliance
Lightning Source LLC
LaVergne TN
LVHW070627170726
843515LV00004B/196

* 9 7 8 3 0 3 6 5 1 1 0 2 3 *